DATA ANALYSIS

for High School Math

With Examples Using
the TI-83 and Similar Calculators

First Edition

by James Bellon

Copyright 2020. All rights reserved. No part of this publication may be reproduced or distributed in any form or by any means, or stored in a database or retrieval system, without the prior written permission of the author.

For questions and support, contact us by email **e3education42@gmail.com**

For virtual tutoring in math and more, visit our website **www.e3education.org**

Contents

Introduction   1

1   Descriptive Statistics   3

    1.1   Collecting Data   3

    Exercises: Collecting Data   10

    1.2   Summarizing Data   11

    Exercises: Summarizing Data   24

    1.3   Measuring Data Sets   26

    Exercises: Measuring Data Sets   35

    1.4   Measures of Relative Standing   37

    Exercises: Measures of Relative Standing   44

    1.5   Data Sets with the TI-83 and Similar Calculators   46

2 Sets and Probability — 53

- 2.1 Sets and Venn Diagrams 53
- Exercises: Sets and Venn Diagrams 62
- 2.2 Probability Basics 64
- Exercises: Probability Basics 71
- 2.3 Counting Rules 72
- Exercises: Counting Rules 76
- 2.4 More Probability 77
- Exercises: More Probability 84

3 Other Topics — 85

- 3.1 The Normal Distribution 85
- Exercises: The Normal Distribution 98
- 3.2 Correlation and Regression 99
- Exercises: Correlation and Regression 111
- 3.3 Converting Units 113
- Exercises: Converting Units 130

4 Solutions — 131

Answers to Try This On Your Own Problems 131

Solutions to Exercises: Sec 1.1 Collecting Data 138

Solutions to Exercises: Sec 1.2 Summarizing Data 140

Solutions to Exercises: Sec 1.3 Measuring Data 144

Solutions to Exercises: Sec 1.4 Measuring of Relative Standing 146

Solutions to Exercises: Sec 2.1 Sets and Venn Diagrams 149

Solutions to Exercises: Sec 2.2 Probability Basics 151

Solutions to Exercises: Sec 2.3 Counting Rules 153

Solutions to Exercises: Sec 2.4 More Probability 155

Solutions to Exercises: Sec 3.1 The Normal Distribution 157

Solutions to Exercises: Sec 3.2 Correlation and Regression 163

Solutions to Exercises: Sec 3.3 Converting Units 167

Standard Normal Table — 170

Index — 172

Introduction

Welcome to Data Analysis! Just what is Data Analysis? The simple answer is "the analyzing of data", but you may want a more detailed explanation than that. Data Analysis is a term commonly used for the collection of fundamental concepts from statistics, probability, graphs, measurements, and converting units.

Memorizing formulas and just being able to do problems are not enough to prepare for advanced math classes and college. In order to succeed in mathematics, it is my belief that students must fully understand the concepts, be able to apply them to real world problem solving, and be able to make sense of the results. Here we will focus on not just the what and when, but also the why, how, and where is it used. Most students who do not succeed in math or dislike math, probably were told: "Just do it the way I told you, get through it quickly and move on. Don't worry about understanding it". I have seen many students with a fear and dislike of math, who have succeeded and actually started to like it, once they are shown the "big picture" and how it all comes together.

The main excuse I hear for failing is "I won't ever need to know this or use this, so why should I learn this?" The benefit of learning mathematics is not necessarily to obtain particular knowledge about certain math topics. Retaining the knowledge is important, if you have to go on to the next level. However, even for those who just need to pass one core math course, there is still plenty to gain. Learning mathematics trains your brain to think logically and develops analytical skills, which can be used for the rest of your life. Some examples include doing your taxes, running a business, managing people or projects, building a tree-house for your kids, etc.

Think of it this way: a professional basketball player is mainly concerned with getting the ball into a basket, so why does he lift weights (the ball is not heavy), run laps (the court is only 94 feet long), and analyze film. Because these things help train him for being better at what he wants to do. The same can be said about learning math. It helps train your brain to be better at dealing with the real world, life, careers, etc. So even if you do not care to know the math itself, learning math can be a vehicle for increasing your brainpower and critical thinking skills.

Here are some tips on how to use this book and what else you can do to succeed in this course. These ideas are not my discovery and can be applied to almost any course.

- Read each section of the book BEFORE you cover it in class. Let your brain mull it over so class time will seem like review or at least let you get a better grip on the material.

- TAKE GOOD NOTES!! I have noticed that many students (especially home-schoolers) do not take notes. Just because you have a textbook, does not mean that you should not write out your thoughts in your own words.

- Do ALL of the exercises, practice makes perfect (or at least closer to perfect).

- Work the problems and think them over BEFORE you look at the solutions.

- Start working on assignments and studying early, this way you avoid "something else came up, I couldn't finish".

- Many test problems will be similar to exercises, but some will combine topics and/or be longer. Prepare very thoroughly for tests.

- Don't be afraid to ask questions. Most teachers/tutors take them seriously. Questions help them to assess where you are BEFORE you're tested.

- Search the internet for one of the many FREE online math help sites.

Chapter 1

Descriptive Statistics

1.1 Collecting Data

If we wish to analyze data, we need to understand what data is. People and things have characteristics which can be observed or measured. A **Variable** is a general type of characteristic (or type of status), which can be different for each person or thing. Some variables are: name, height, color, texture, mood, wingspan, density, anxiety level, etc. **Data** is the collection of all observations for a particular variable or variables, from one or more people or things.

The branch of mathematics that covers the methods and procedures in analyzing data is called **Statistics**. Statistics includes methods for planning studies and experiments, obtaining data, and then organizing, summarizing, presenting, analyzing, interpreting, and drawing conclusions based on data. Below are some important terminology we will be using in data analysis.

A **Population** is the collection of all individuals or items under consideration in a study.

A **Census** is information (data) obtained from the entire population.

In reality, most large censuses (such as the US national census) are an attempt to collect from the entire population, but being so large, some data is never collected. There are just some people who do not wish to be found and others who are too busy to report their data. The results are often adjusted to be a good approximation of the population information. The US census is so large and takes so much time, money, and staff, that it is done only every ten years.

In many cases, it is usually easier and sufficient to collect data from only some of the elements in the population. A **Sample** is the part of a population from which information is actually collected.

A **Parameter** is a numerical measurement describing some characteristic of a population. Examples: The average starting salary of elementary school teachers in Georgia is $33,673. The average for the whole United States is $35,763.

A **Statistic** is a numerical measurement describing some characteristic of a sample. Example: A survey of ten job postings for elementary school teachers in the Atlanta area, had an average starting salary of $38,541.

Try this on your own: For the following scenario, describe the population of interest, describe the sample, state the parameter of interest, and the statistic that was calculated.

A farm wants to track the weight gain of their chickens after they switched to a new feed. The farm has over 10,000 chickens. They isolated 200 chickens and weighed them before the switch, then every week for the next 10 weeks. At the end of 10 weeks, the 200 isolated chickens gained an average of 1.2 pounds.

There are two main types of variables. In this book, we will focus mainly on data from numerical variables, since you can perform calculations with numbers more easily.

Qualitative Variables are variables which have values that are words, symbols, or categories. They can also be numbers that have no absolute measure, order or units. Examples: gender, job title, letter grade, phone number, numbers on football jerseys, etc.

Quantitative Variables are variables which have values that are numerical values with a specific order and units. They represent counts or measures. Examples: height, weight, temperature, number of siblings, hours of sleep, etc.

Quantitative variables can be further broken down into two different types. **Discrete** variables have possible values that form a finite set of numbers or values (typically a count). **Continuous** variables have possible values that form an interval of numbers and can be almost anything between (typically a measurement).

Examples: Number of siblings is discrete since the only possible values are 0, 1, 2, 3, ... up to some realistic maximum. Adult height is continuous, since it can be ANY value between 21 inches (shortest ever recorded) to 8 feet 11 inches (tallest ever recorded), including fractional measurements.

Often the purpose of a statistical study is to investigate whether a relationship exists between 2 characteristics, such as smoking and lung cancer, age and income, or stock price and company revenues. We distinguish between 2 types of procedures.

In an **Observational study**, researchers observe characteristics and take measurements but don't attempt to modify the subjects being studied. An example would be a study of animals by simply hiding and watching what they do.

In a **Designed experiment**, researchers impose treatments and controls and then observe characteristics and take measurements. An Experiment can help establish causes. The subjects are called experimental units. Experiments have **Treatments**, which are the variables that are controlled and changed in order to test the effects. Drug trials are experiments. Different drugs and different doses are given to different groups to see which one is the best.

In experiments, they often use a **Placebo**, which is an inert substance that is used in place of a treatment. It has no direct effect, unknown to the subject. The purpose of a placebo is to eliminate psychological effects of the subjects. In drug tests, they will have a control group who do not get any of the drug. If they know they are not getting drugs, they might feel negative and stressed about not being helped, so the researchers will typically give them a fake pill.

One problem that arises in studies is **Bias** . Bias is when the results systematically favor certain outcomes. To eliminate bias, treatments are assigned **Double Blind** where neither the subjects nor experimenters know who receives which treatment and who gets the placebo until the results are recorded.

How do we make sure the design is truly random? Some methods for obtaining a random sample are to pick cards, roll dice, pick out of a hat, use random numbers from software. The method that is most useful in schools, is use of a random number generator from our calculators.

There are several ways to pick the subjects that will be in a sample. **Probability sampling** is using a random method to select the sample from population. This is better than human judgment, but no guarantee of getting a perfect sample.

Simple random sampling is a sampling procedure for which each possible sample of a given size is equally likely. If the sample is chosen with replacement, then each member of population can be selected more than once. Without replacement means each member of population can be selected only once. We will assume simple random sampling without replacement (unless otherwise specified).

Systematic Random Sampling is a sampling method that follows a system (pattern) and so is not random, but is easy especially for computers. The steps are as follows:

1. Divide the population size, N, by the desired sample size n and always round DOWN to the nearest whole number. We will refer to this number as m.

2. Use random number generator to get a number between 1 and m. We will refer to this number as k.

3. Then systematically chose the members of the sample from the population list starting at number k, and continue with k+m, k+2m, etc., until you obtain the sample size n.

Example: if we want to select a sample of $n = 15$ people from a population of $N = 1400$ people, then $\frac{N}{n} = \frac{1400}{15} = 93.33$, so $m = 93$. We can get a random number from a function on the TI83 calculator. Press the $\boxed{\text{MATH}}$ button, scroll right to the PRB menu, scroll down and select the 'randInt' function and type (1,93) and hit enter. When I did this, I got $k = 17$. Realize that each time we do this it is random, so we may get a different number between 1 and 93. Now the sample members will be from population list as follows: 17, 110, 203, 296, 389, etc., ending with the 15th selection 1319. Notice that if we add 93 again by mistake, we will be over 1400 and there will not be anyone to choose.

In **Cluster Sampling**, the population is divided into clusters (usually based on location). Randomly choose a cluster and use the members to get the sample size. If you need more to complete the sample, randomly choose another cluster and use part or all as needed.

The method that usually has least amount of bias is **Stratified Sampling**, where the population is divided into sub-populations called strata (one is called a stratum). Members within a particular stratum should have common characteristics relative to the statistical study. Then a simple random sample is taken from each stratum in close proportion to the size of the stratum. The strata samples are combined to obtain the overall sample.

Example: If we wish to choose a stratified sample of 11 people, from a population of 78 males and 42 females, then the male sample should be about $\frac{78}{120} = 0.65 = 65\%$. For 11 people, 65% of 11 is 7.15, so randomly choose 7 of the 78 males, then the other 4 people will be randomly chosen from the females.

Convenience Sampling is when we just select the easiest elements to be in the sample. For example, if you need 10 people for a survey, then ask the first ten people you come in contact with. This can often lead to extreme bias. The ten people might be related or friends, and have similar opinions.

Try this on your own: What sampling methods would each description below be classified as?

1. A teacher selected a sample of students by selecting one row and picking all students in that row.

2. A researcher was conducting a survey where they selected a sample by going to every tenth neighborhood and surveying every tenth home from those neighborhoods

Whether conducting statistical analysis of data that we have collected, or analyzing a statistical analysis done by someone else, we should not rely on blind acceptance of mathematical calculation. We should consider these factors:

1. Context of the data: What do the values represent? Why were they collected? An understanding of the context will directly affect the statistical procedure used.

2. Source of the data: Is the source objective or biased? Is there something to gain or lose by distorting results? Be vigilant and skeptical of studies from sources that may be biased, such as a nutrition study done by a fast food company.

3. Sampling method: Is the method chosen appropriate and help eliminate bias? Voluntary response(self-chosen) samples often have bias (those with strong opinions are more

likely to participate). These sample results are not necessarily valid. Other methods are more likely to produce good results.

4. <u>Conclusions</u>: Make statements that are clear to those without an understanding of statistics and its terminology. Avoid making statements not justified by the statistical analysis.

5. <u>Practical implications</u>: State practical implications of the results. The results may be valid and significant yet there may be NO practical significance. Does anyone even care about it? Common sense might suggest that the finding does not make enough of a difference to justify its use or to be practical.

6. Consider the likelihood of getting the results by chance. If results could easily occur by chance, then they are not statistically significant (you did not justify anything). If the likelihood of getting the results by chance is so small, then the results are statistically significant (you found strong evidence).

7. It is important to carefully plan the study and know what you are trying to show before you do any work. Improper planning can result in a poor or incomplete study. Make sure you have enough resources (time, money, people, supplies) to complete your study and report your results in a professional manner.

Exercises: Collecting Data

Solutions appear at the end of this textbook.

1. Identify the population, sample, parameters, and statistics for the following situation. A textbook company wants to know the average price of homeschool science textbooks in the United States. They obtain a list of 15 science books and compute the average price of the 15 books is $52.

2. A teenager put the following information on his myspace page. What are the variables, what are their values, which variables are qualitative, which are quantitative? Name: Sean Higgins, Ht: 5ft.10in., Wt: 185 lbs., Eyes: Green, Hair: Red, Page-hits: 142

3. Use systematic sampling to select 7 people out of a group of 6700. Assume the random value of $k = 121$ (so we all have the same results). State the place numbers of the selected sample.

4. Does the following describe an experiment or just an observational study? Explain why. Sarah was on a field trip for her science class. At the beach, she saw a sand castle with 2 crabs crawling inside it. She timed the crabs to see how long they took to find their way out.

5. Which type of sampling best fits the following? Explain why. DJ Paulie D is looking for some songs to be the background beats for his new mix. He sorts his ipod collection into four categories: Rap, Instrumental, Dance, and Alternative. Then he randomly picks 5 songs from each category and listens to the beats.

6. What is a census? The US government does a census every ten years. Why don't they do one every year? Is the US census an actual census? Why or why not?

7. Which sampling methods tend to have bias? Explain how they have bias.

8. Why do experimenters use placebos? How do they use them?

1.2 Summarizing Data

After data is collected, it is a good idea to summarize and display the data in ways that show the important characteristics. One of the best ways is to create a **Distribution** of each variable, which is information that tells us what values the variable takes on and how often. Large sets of data are often grouped according to **Classes**, which are categories or groupings (quantitative data is in consecutive intervals). The three guidelines for grouping data into classes are:

1. Small number of classes to be effective, but enough to show differences.

2. Each observation must belong to only one class.

3. Whenever feasible, classes should have same width.

Then compute the **Frequency** of each class, which is the number of observations that fall into a class (count). A listing of all classes of the data and their frequencies is called a **Frequency distribution**.

When data sets are different sizes, it is hard to compare them. A good way to compare is to compute **Relative Frequency**, which is the ratio of the frequency of a class to the total number of observations. A listing of all classes and their relative frequencies is called a **Relative Frequency distribution**. Most distributions show frequencies as well as relative frequencies.

Example: Bradley worked a summer job to earn money for college. His weekly hours over a 12 week period were 25, 32, 36, 32, 18, 28, 30, 36, 12, 16, 35, 36. We can group the hours into 3 classes in various ways, but one simple choice would be 10-19, 20-29, and 30-39. The Distribution would be as follows:

Hours	Frequency	Relative Frequency
10-19	3	$\frac{3}{12} = 0.25 = 25\%$
20-29	2	$\frac{2}{12} = 0.167 = 17\%$
30-39	7	$\frac{7}{12} = 0.583 = 58\%$
Total	12	100%

Notice that the frequencies add up to 12, and there are 12 weeks of data. Also the relative frequencies add up to 100%. These must always happen, or the distribution was done incorrectly. If the relative frequencies are rounded, then total percentage may be slightly off, between 99-101%. Any larger difference is not valid.

As the old saying goes, a picture is worth a thousand words. Data summaries can come in pictures or graphs. Here are some of the typical types of graphs to display distributions. They can give us a quick overview of the big picture and the characteristics of the data.

A **Frequency Histogram** is a graph that displays the classes on the horizontal axis and the frequencies on the vertical axis. It consists of vertical bars, whose height is equal to the frequency of the class(interval). The bars are drawn next to each other (without gaps), since they encompass the range of the data in numerical order. A Histogram is only for quantitative data, not qualitative.

A **Relative Frequency Histogram** is the same as a frequency histogram, except it uses relative frequencies for the vertical axis and the bar heights.

Example: The frequency and relative frequency histograms for Bradley's summer job data are shown below.

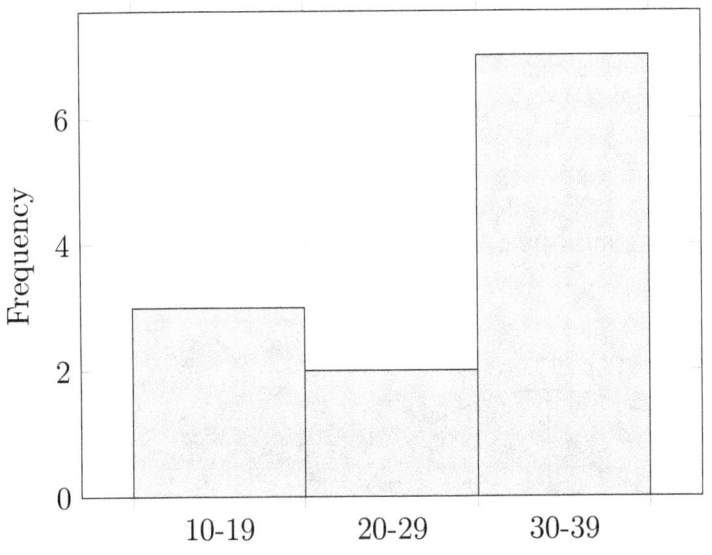

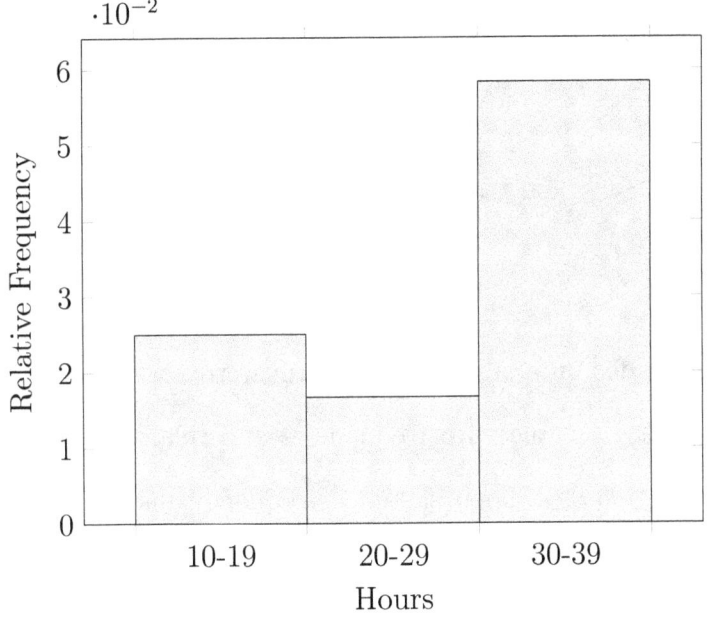

Notice that these graphs are the same shape, this is because the relative frequencies are based on the frequencies, so the same relationships between the classes are maintained.

A **Pie Chart** is a disk (circle) divided into pie-shaped pieces proportional to the relative frequencies. A pie chart should be labeled well, with class and the relative frequency for each slice. If a slice is very small, then the labels can go outside with an arrow pointing to the corresponding slice. The preferred way to sketch a pie chart is to start slices at 12 o'clock and rotate clockwise.

Example: The pie chart for Bradley's summer job data is shown below.

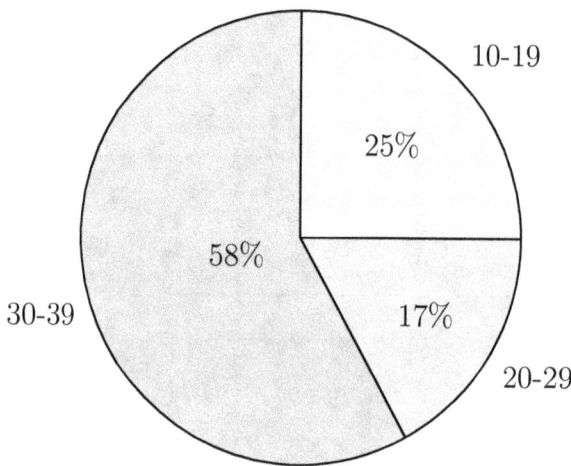

**Try this on your own: The grades on a science final exam were 75, 83, 96, 82, 90, 78, 60, 76, 82, 71, 92, 86, 83, 88. Create a table with frequencies and relative frequencies using the intervals 60-69, 70-79, 80-89, 90-99. Then sketch a frequency histogram and a relative frequency pie chart.

Histograms are for quantitative data. There is a similar graph for qualitative data (categories), called a **Bar Graph**. In a bar graph, the width of the bars is arbitrary and the bars are not connected. Can show frequency or relative frequency. A **Pareto chart** is a specific type of bar graph where the classes are reordered so that the bars are in size order.

Example: The US Men's olympic hockey teams have played in 21 olympic games, winning 11 medals (2 gold, 8 silver, 1 bronze). Below are a frequency bar graph (ordered from worst to best finishes) and a relative frequency pareto chart (ordered from highest to lowest).

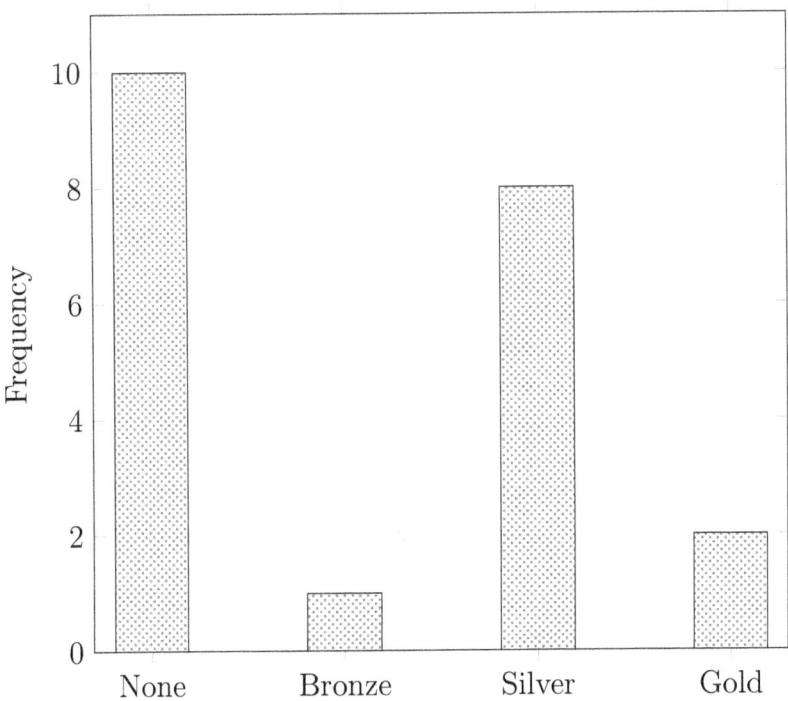

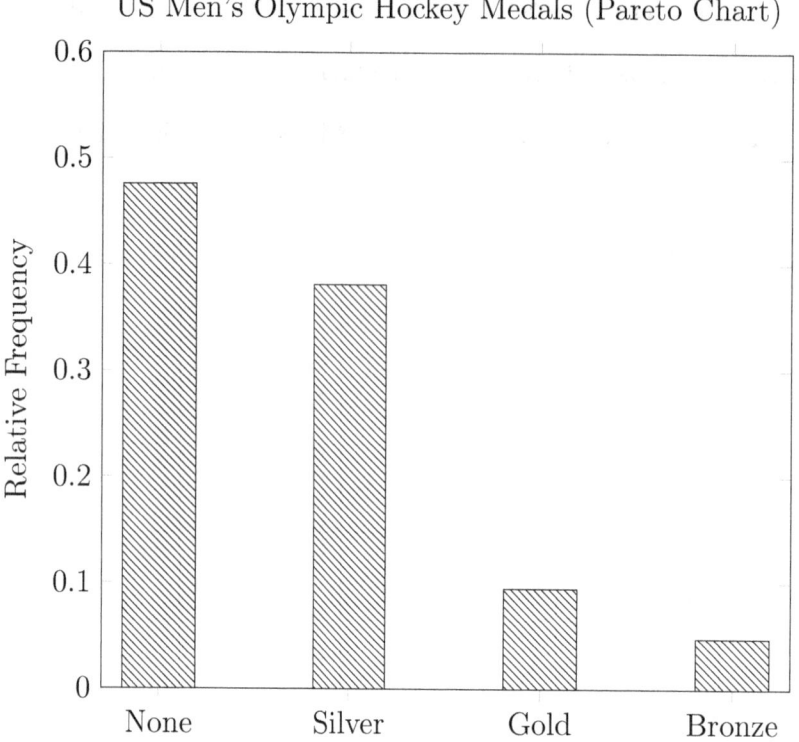

Graphs can help us get an overall view of the data set. When looking at a graph, pay attention to the following:

- Center: where is the middle of the graph, and the highest point.

- Spread: how are the parts of the graph spread out from each other?

- Shape: what shape does the graph have? Bell shape, straight across, repeatedly up and down, random?

- Symmetry: graph can be split in half with two mirror image parts, almost equal amount on both sides. Graph that extends more out to left is **Left-skewed**. Graph that extends more out to right is **Right-skewed**.

- Outliers: are data values (small parts of the graph) that are far from the other data (parts).

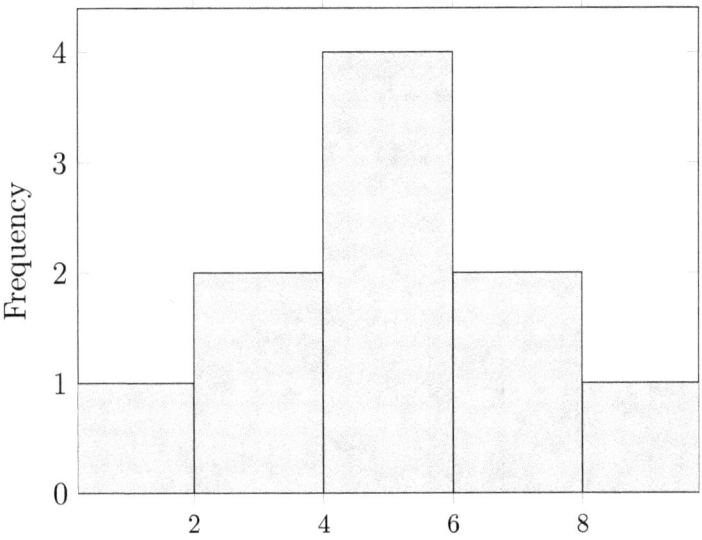

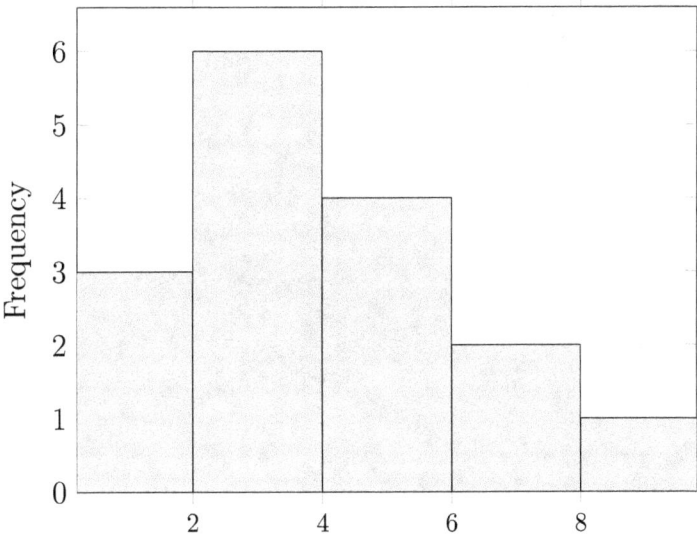

17

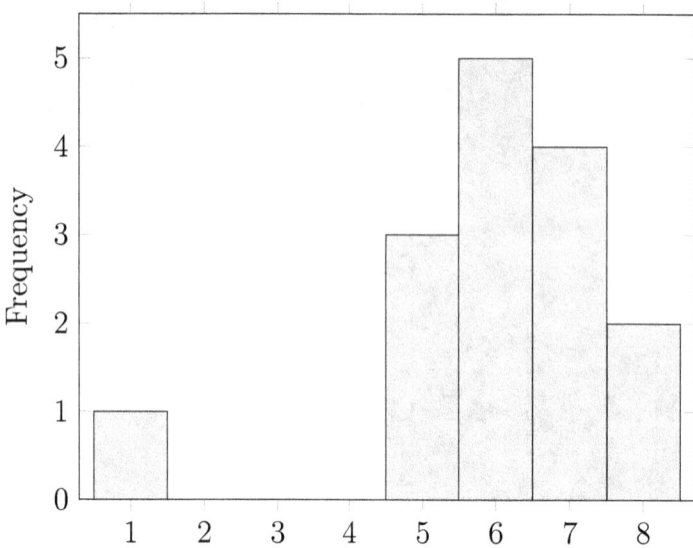

Example: what are the characteristics of the following graph? What does is suggest about the data?

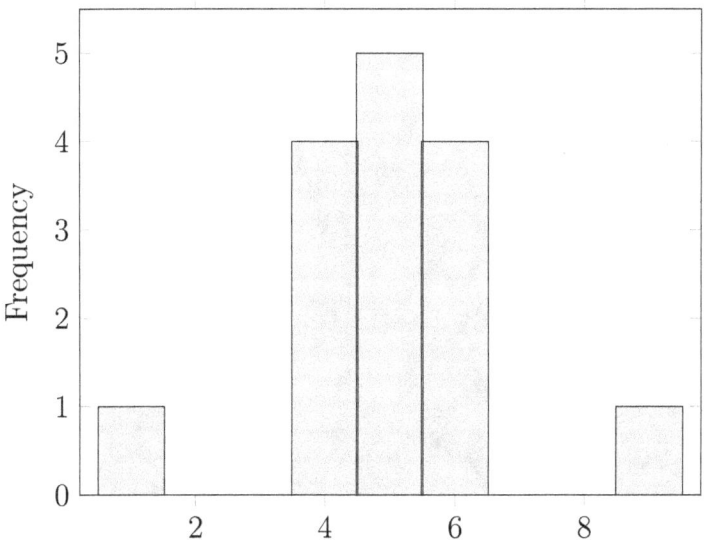

Solution: The center of the graph is at data value 5, also the highest point is at 5. Most of the data in concentrated near the middle (from 4 to 6) with outliers on both extremes (only one value at 1 and one value at 9). It is not spread out very much. The graph is symmetric, mirror images on either side of 5. It is somewhat bell-shaped.

Another type of simple graph that plots the values of a single variable over time, is called a **Time Series**. The horizontal axis is in time units and the vertical axis is scaled for the values of the variable being graphed. It has values plotted as points and connected by lines, although the lines themselves do not represent data, they are just to show the pattern of the points. The vertical scale should start at zero, unless all of the data is very large, then it is better to start higher.

Example: Here is a time series graph for the Dow Jones stock market closing average for the first business day of each month of 2014.

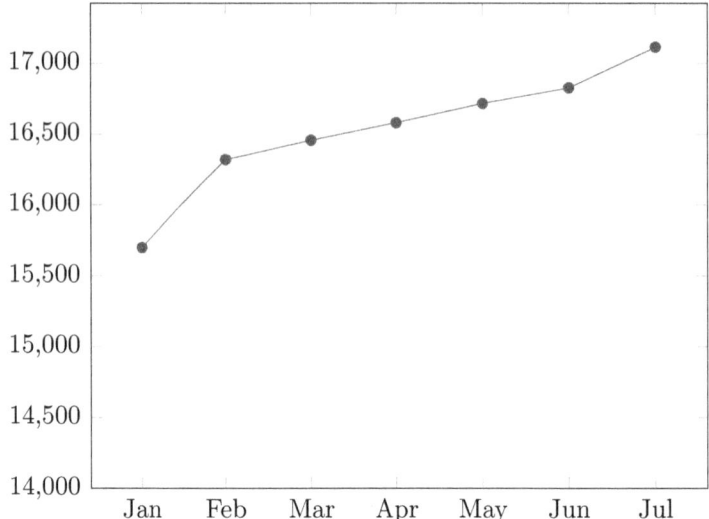

Dow Jones closing value: 1st business day of month, 2014

From this graph, we can see a general upward trend (increasing over time), with the sharpest increase from January to February. Notice the vertical axis for the price, does not start at zero. This is because the values are all very large, and starting at zero would collapse it into a small area at the top. It would be hard to see any details.

Beware of misleading or bad graphs!! Any data can be shown accurately but with different graphs and seem like it is showing very different results. There are several common ways that graphs can be misleading.

- Starting the vertical axis (values or frequencies) above zero. This chops the bars down and exaggerates the differences. The only exception is for line graphs (time series) with very large values for all of the data, to avoid having the graph squished into a small area.

- Using uneven scales on the axes.

- Using multiple dimensions when the data is just one dimension (using area or volume, when data is only the height for the bars).

- Unclear labelling.

- Too many cosmetic enhancements. This makes the graph hard to read, it is too busy!

- Poor choice of grouping.

Here is an example of how to create a good or bad graph with the same data. The first graph is bad because it starts the vertical axis at 100, which makes the increase look very steep. It is also not labelled. The second graph shows a better view of the data starting at zero and labelled.

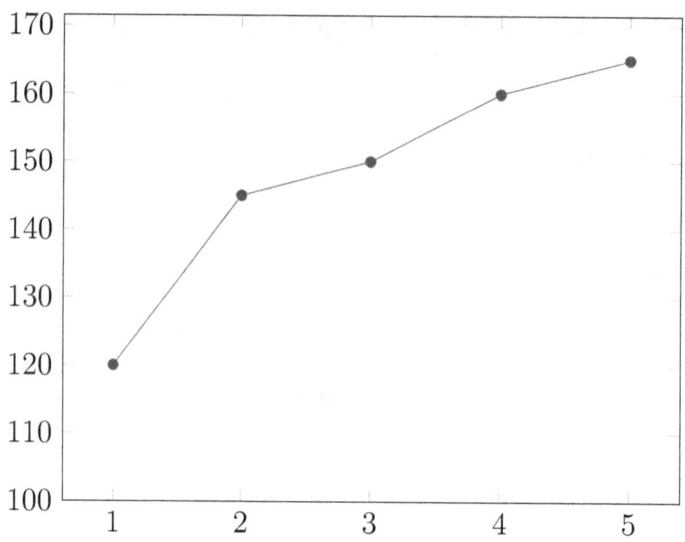

Notice the increase from 1 to 2, looks very steep above, due to the vertical axis being chopped down. Below, the axis starts at zero and shows a less drastic increase. Also the graph below is easier to read with labels.

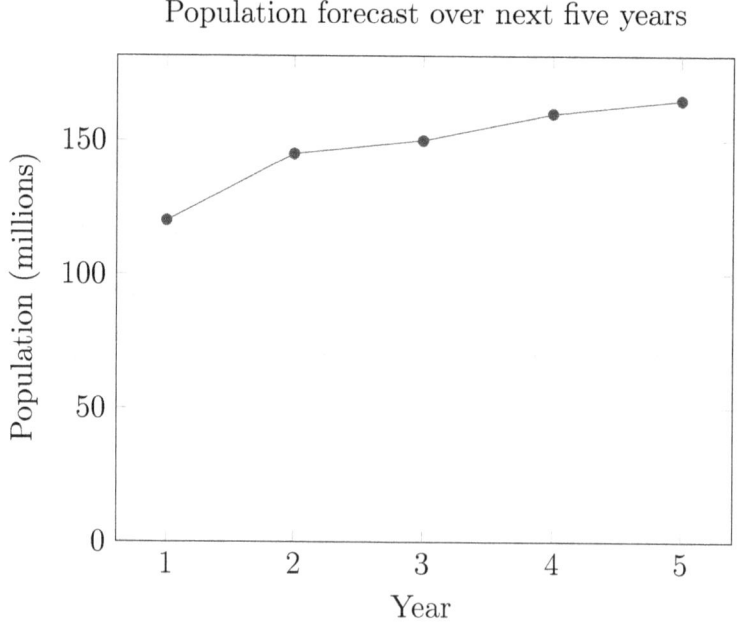

**Try this on your own: What are the characteristics of the following graph? Examine the spread, symmetry, and outliers.

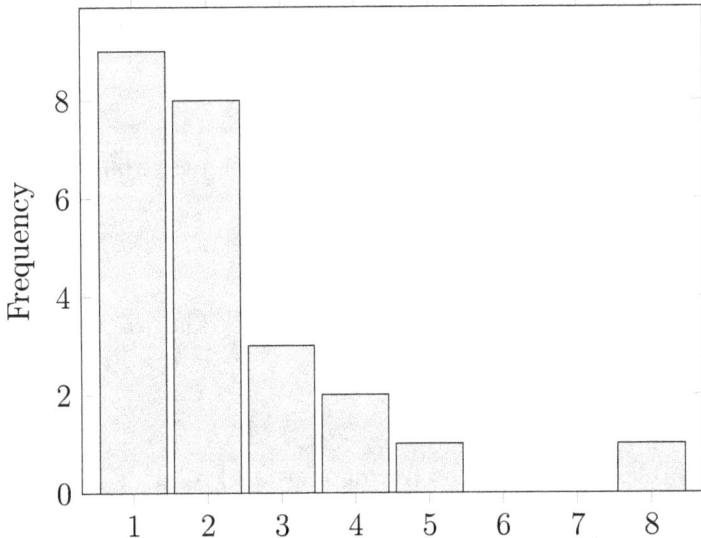

Exercises: Summarizing Data

Solutions appear at the end of this textbook.

1. Why should we use only a small number of groups? If a data set has 1000 values, why not use 100 groups?

2. When computing relative frequencies, how do we know that we have calculated them correctly?

3. What is the difference between a bar graph and a pareto chart? Can a pareto chart be done from quantitative data? Why or why not?

4. Group the following data into classes, calculate the frequencies and relative frequencies. Convert relative frequencies to nearest whole percent. A Farmer kept a log for one month of which days it rained. Here is the order. Tuesday, Saturday, Sunday, Tuesday, Friday, Sunday, Wednesday, Sunday, Friday, Tuesday

5. Make a pie-chart and a bar graph for the distribution you created from the farmer's observations in the previous exercise.

6. Group the following data into classes, calculate the frequencies and relative frequencies. Convert relative frequencies to nearest tenth of a percent. The grades on a history test were as follows: 67, 72, 99, 100, 82, 83, 94, 90, 80, 85, 85, 77, 48, 88, 75, 50, 75, 82, and 95. Use Classes of F, D, C, B, and A.

7. Make a pie-chart and a histogram for the distribution you created from the history grades in the previous exercise.

8. Describe the characteristics of the following graph (center, spread, shape, symmetry, outliers).

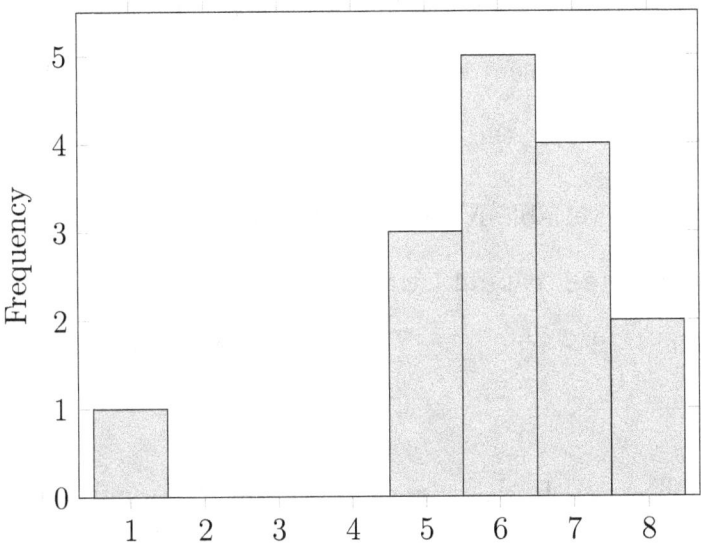

9. Describe two issues that are bad or misleading about the following graph.

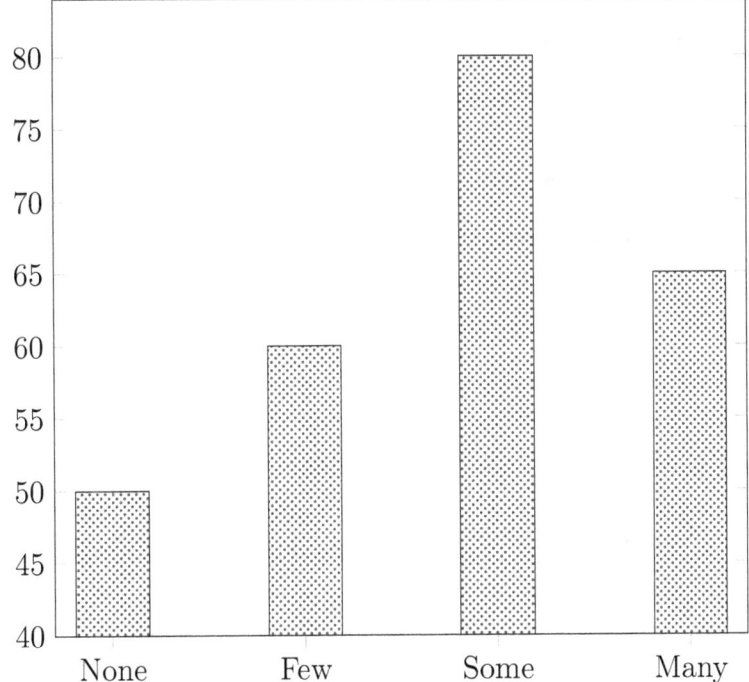

1.3 Measuring Data Sets

In order to use mathematics to measure data sets, there will be several formulas that we will use. We will typically use letters to represent variables (typically X or Y or a letter that has some relevant meaning like S to represent the variable salary).

Particular values (observations) of a variable X can be denoted by subscripts x_1, x_2, etc. For summation of values, we use the Greek capital letter Sigma Σ, with a variable next to it to show which one is being summed. For example: $\sum x$ stands for the sum of the values of the variable X.

Descriptive measures are numbers used to describe data sets (average, min, max, etc.). They fall into two main categories: Measures of Center and Measures of Variation. We will use a common rule for rounding here.

The **Round-Off Rule**: Round all calculations to one more decimal place than is present in the data. Round only the final answer, not the steps along the way. For example, if the data set has values that are go out to 2 decimal places, then all calculated statistics should be reported showing 3 decimal places.

Measures of Center (or measures of central tendency) are descriptive measures that indicate where the center or most typical value of a data set lies. Some of the specific measures of center are shown below.

The **Mean** is sum of all the values of the observations, divided by the number of observations. The mean is also more commonly just called average .

The Sample mean is represented by the symbol $\bar{x}$, which is called 'x-bar': $\bar{x} = \frac{\sum x}{n}$, where n is the number of values in the sample.

The Population mean is represented by the symbol μ, which is called 'mu': $\mu = \frac{\sum x}{N}$, Where N is the number of values in the population.

Both of these are the same procedure and give the average of the values, the only difference is where the values come from, a sample, or the whole population.

Example: In the previous section, we looked at graphs for Bradley's weekly hours at a summer job: 25, 32, 36, 32, 18, 28, 30, 36, 12, 16, 35, 36. Find the mean (average) hours he worked in a week.

Solution: We add up the values and divide by 12 (the count of how many values). Since this is the entire set of his summer job hours, it is a population mean. $\mu = \frac{\sum x}{N} = \frac{336}{12} = 28$, which we will report, according to the round off rule, as 28.0 hours worked in an average week.

The **Median** is the value that divides the bottom 50% of data from top 50%. It is the middle value when the values are placed in size order. To find the median, first arrange the data in increasing order. If there are an odd number of observations, the median is the middle value in order. If there are an even number of observations, the median is the average of two middle values in order.

Example: Find the median of Bradley's summer weekly hours.

Solution: First we must put the values in order from lowest to highest: 12, 16, 18, 25, 28, 30, 32, 32, 35, 36, 36, 36. There are two values in the middle, 30 and 32, with five values below and above. The average of the two middle values is 31, which is the median. Notice that the median is not one of the original data values here. According to the round-off rule, we will report the median as 31.0 hours worked in a week. Half of the data is below this and other half above this.

The **Mode** is the value that has the most number of observations (frequency), but must occur more than once. There can be multiple modes (a tie for the most often).

Example: Find the mode of Bradley's summer weekly hours.

Solution: Having the values in order makes this easier: 12, 16, 18, 25, 28, 30, 32, 32, 35, 36, 36, 36. The values that occurs the most often is 36, which is the only mode here.

Here are some cautions about measures of center. The mean is sensitive to extreme values. If a company has 20 workers making $15,000 each and the owner makes $500,000, then the mean would be $38,095. This mean does not give a complete picture of the company salaries. Nobody makes close to that value, everyone except the owner is way below average.

When dealing with salaries or prices, the median is often a better measure of the data set. The median for the company would be $15,000 and a better representation of what potential employees could expect to be paid.

Most of the time, data that is left-skewed, will have a mean that is less than the median. For right-skewed, the mean is greater than the median. This is because the skewed data out to the extreme, pulls the mean closer to that extreme, but the middle values are still in place so the median is not affected.

If a large data set has already been grouped and you have only the frequency distribution (but not the actual data), the average can be estimated using the midpoint of each class (group) as the estimate of the typical value in each class and multiplying that by the frequency of that class. The formula is: $\frac{\sum \hat{x} f}{\sum f}$, where $\hat{x} f$ is the product of each class midpoint $\hat{x}$, times the class frequency, f.

Example: Estimate the mean of the following frequency distribution.

Class	1-8	9-16	17-24	25-32	33-40
Frequency	3	5	7	2	1

Solution: The midpoints of the classes ($\hat{x}$) are: $4.5, 12.5, 20.5, 28.5, 36.5$. Then the formula would be $\frac{\sum \hat{x} f}{\sum f} = \frac{4.5(3)+12.5(5)+20.5(7)+28.5(2)+36.5(1)}{3+5+7+2+1} = \frac{313}{18} = 17.3889$

According to the round-off rule, we should report this as 17.4. Let's think about the estimate of 17.4, does it make sense? Notice in the frequency distribution, that the class 17-24 occurs the most, with more in the lower classes than in the upper classes. Then the estimate of 17.4, in the beginning of the class 17-24, makes sense.

Another special case of finding average is when we know the values, but they are not all equally weighted. This idea is know as a **Weighted Mean**. The formula is $\bar{x} = \frac{\sum xw}{\sum w}$, where xw is the product of each data value x multiplied by its weight w. The weight could be a dollar amount or a credit amount (as in college classes), or other amounts.

Example: If an investment earns 6% interest on $1,000 and 4% interest on $250, what is the average interest rate for the entire investment.

Solution: here we are looking for the average interest rate, so the data values are the interest rates. The dollar amounts are the weights. More money, means more weight for its corresponding interest rate. Therefore, the average rate will be closer to 6%, since it corresponds to the biggest dollar amount invested. $\bar{x} = \frac{6(1000)+4(250)}{1000+250} = \frac{7000}{1250} = 5.6$, so the average interest rate on the entire investment is 5.6%.

Example: Rachel's class is graded in the following manner: Test average counts 30%, homework average counts 25%, participation counts 10%, project counts 15%, and final exam counts 20%. What is her overall grade average if she has the following grades:
Test avg 86, HW avg 94, part. 100, project 80, final 82.

Solution: here we are looking for the average grade, so the data values are grades. The percentages are the weights. Test average has the most weight (30%).

Class Avg $= \frac{86(30)+94(25)+100(10)+80(15)+82(20)}{30+25+10+15+20} = \frac{8770}{100} = 87.7$, so her overall grade in the class is about 88, a high B. Almost an A, but not quite.

Academic Grade Point Average (or GPA) is a very common concept, but one that most people don't understand how to compute. Letter grades earned in classes are assigned a numerical value called quality points, ranging from 0 to 4. Typical grade scale is $A = 4.0$, $B = 3.0$, $C = 2.0$, $D = 1.0$, and $F = 0$. Some schools have grades in between with a $\pm$, with fractional values. The weights of the grade points are the credit hours for the classes. A longer class for 4 credit hours has more weight than a short elective for 2 credits.

Example: Compute the overall semester GPA for a college student who earned the following grades. Use the standard letter points shown above, assume there are no $\pm$ grades in between.

Course	Grade	Credits
Physics	B=3.0	4
English	C=2.0	3
Math	C=2.0	3
Study Skills	B=3.0	1
History	A=4.0	3

Solution: here we are looking for the average grade points, so the data values are grade (quality) points. The credit amounts are the weights.

$$GPA = \frac{3.0(4) + 2.0(3) + 2.0(3) + 3.0(1) + 4.0(3)}{4+3+3+1+3} = \frac{39}{14} = 2.79, \text{ which is a high } C,$$

close to an overall B. Not the greatest, but it is passing.

Measures of Variation (or measures of spread) are descriptive measures that indicate how much variation is in the data or how spread out the data values are from each other.

The **Minimum** (Min) is the lowest value in the data set.

The **Maximum** (Max) is the highest value in the data set.

The **Range** is the difference between Min and Max: ($Range = Max - Min$). To give a more detailed range, some studies will just say the range goes from MIN to MAX.

Example: Find the min, max, and range of Bradley's summer weekly hours.

Solution: Having the values in order makes this easier as well: 12, 16, 18, 25, 28, 30, 32, 32, 35, 36, 36, 36. Here $min = 12$, $max = 36$, and $Range = 36 - 12 = 24$ hours. We could also say the range is from 12 to 36 hours.

Perhaps the hardest measure to compute (without using special function on calculators or computers) is the next measure of spread.

Standard Deviation is the measure of how far, on average, the data is from the mean. Another related measure, is the **Variance** which is standard deviation squared. It will not be used much here, except as a step in the computation of standard deviation.

The standard deviation and variance for a SAMPLE are calculated by the following symbols and formulas:

Variance: $s^2 = \dfrac{\sum (x - \bar{x})^2}{n - 1}$ $\qquad$ Standard deviation: $s = \sqrt{\dfrac{\sum (x - \bar{x})^2}{n - 1}}$

The variance and standard deviation for a POPULATION are calculated by the following symbols and formulas:

Variance: $\sigma^2 = \dfrac{\sum (x - \mu)^2}{N}$ $\qquad$ Standard deviation: $\sigma = \sqrt{\dfrac{\sum (x - \mu)^2}{N}}$

Example: Find the standard deviation of Bradley's summer weekly hours.

Solution: To work out standard deviation yourself, it is very important to work out each step carefully and organized. The data set is the population of the entire summer hours for Bradley, so we will use the second set of formulas. First we use the mean $\mu = 28$ and the total count $N = 12$ from the previous examples. Then it helps to make a table like the one below. For repeated values, the calculation is only shown once (one row), but used multiple times in the formula.

x (hours)	$x - \mu$	$(x - \mu)^2$	Count
12	$12 - 28 = -16$	$(-16)^2 = 256$	once
16	$16 - 28 = -12$	$(-12)^2 = 144$	once
18	$18 - 28 = -10$	$(-10)^2 = 100$	once
25	$25 - 28 = -3$	$(-3)^2 = 9$	once
28	$28 - 28 = 0$	$0^2 = 0$	once
30	$30 - 28 = 2$	$2^2 = 4$	once
32	$32 - 28 = 4$	$4^2 = 16$	used twice
35	$35 - 28 = 7$	$7^2 = 49$	once
36	$36 - 28 = 8$	$8^2 = 64$	used 3 times
Sum		642	

The standard deviation, $\sigma = \sqrt{\dfrac{\sum (x - \mu)^2}{N}} = \sqrt{\dfrac{642}{12}} = \sqrt{53.5} = 7.31437$

Using the round off rule, $\sigma = 7.3$ hours. Computing the standard deviation can be done very quickly and easily using special calculator functions or computers, but interpreting the meaning of it requires human understanding. In this example, it means that Bradley's weekly hours are somewhat spread out, on average his weekly hours are 7.3 hours away from the mean of 28 hours. Sometimes they were close to 28 (one week exactly), and sometimes farther away (as much as 16 hours below).

If another employee worked with Bradley, but had a standard deviation of only 3 hours, then that would mean that the other employee worked weekly hours that were much more consistent, working about the same each week, and not as spread out as Bradley's hours were.

If we wish to compare two data sets, to figure out which is spread out more, there are two cases to consider. First, if the data sets are from similar variables with similar sizes, then we can directly compare the standard deviations, since they are the same units. As an example, we already discussed Bradley's hours versus his fellow employee.

The other case is when comparing two very different data sets. For that we will use a special measure called the **Coefficient of Variation** (or CV). It is equal to the standard deviation divided by the mean, converted into a percent. It has no units, it is only a ratio as percent. The formulas are slightly different, depending upon the data set being from a sample or a population. The CV states how big the standard deviation is, relative to the average size of the data.

For a sample: $CV = \dfrac{s}{\bar{x}} \times 100\%$ For a population: $CV = \dfrac{\sigma}{\mu} \times 100\%$

Example: Which data set is more spread out, the weight of elephants in a herd: $s = 1,175$ pounds and $\bar{x} = 12,342$ pounds, or the price of regular unleaded gasoline in a US: $s = \$0.26$ and $\bar{x} = \$3.73$?

Solution: For the elephant weights, $CV = \dfrac{1,175}{12,342} \times 100\% = 9.5\%$. For the gas prices, $CV = \dfrac{0.26}{3.73} \times 100\% = 7.0\%$. The weights of elephants are a more spread out set of data than US gas prices. This is NOT because the values are larger. Another set of large values could have a lower CV than gas prices.

Try this on your own: The grades for a sample of a science final exam were 75, 83, 96, 82, 90, 78, 60, 76, 82, 71, 92, 86, 83, 88. Calculate the mean, median, mode, range and standard deviation.

Exercises: Measuring Data Sets

Solutions appear at the end of this textbook.

1. For the following data, compute the mean, median, and mode. Use the round-off rule. The grades on a history test were as follows: 67, 72, 99, 100, 82, 83, 94, 90, 80, 85, 85, 77, 48, 88, 75, 50, 75, 82, and 95

2. For the following data, compute the min, max, range, variance and standard deviation. One of the hottest selling concerts last summer was the Honda Civic Tour featuring Maroon 5, Kelly Clarkson, and Rozzi Crane. A sample of ticket prices for the Atlanta Lakewood show on August 1st were $44, $74, $94, and $116.

3. Explain what the standard deviation from the previous exercise tells you about the ticket prices.

4. How are the mean, median and mode affected by extreme values?

5. If a set of data is not known to be from a sample or a population, then there are two possible standard deviation formulas to use. Explain why the sample standard deviation is always larger than the population standard deviation.

6. College GPA is a weighted average of the grades earned in all of your courses. Letter grades are equated to a numeric value called quality points. The weights are the credit hours earned for the course. On a typical A-F system with no plus/minus, the points are $A = 4.0$, $B = 3.0$, $C = 2.0$, $D = 1.0$, and $F = 0$. If a student had the following grades for their first semester, what would their GPA be for that semester? Grade B in College Algebra(3 credits), Grade B in Chemistry(4 credits), Grade A in Phys Ed(2 credits), Grade C in Writing(3 credits), Grade A in Economics(3 credits).

7. A fitness company did a study and found the following statistics for 1000 women in the Atlanta area. Mean weight = 162 pounds, median weight = 141 pounds, standard deviation = 45 pounds. What do these statistics suggest about the distribution of women's weights?

8. Estimate the mean of the following frequency distribution.

Class	10-14	15-19	20-24	25-29	30-34	35-39
Frequency	12	5	7	2	6	3

9. Which data set is more spread out, the shot put throws for a high school track team: $s = 5.5$ feet and $\bar{x} = 38$ feet, or the gymnastics scores for a college team: $s = 1.4$ points and $\bar{x} = 8.45$ points?

1.4 Measures of Relative Standing

In the last section, we looked at measures which described the data set as a whole. In this section we will look at measures that describe how particular data values compare to each other, called **Measures of Relative Standing**.

A **z-score** (or standardized score) is the number of standard deviations that a given value is above or below its mean. Whenever a value is below the mean, its corresponding z-score will be negative. Usual values are z-scores from -2 to $+2$. Unusual values are z-scores outside this range. More than 3 standard deviations away from the mean is very unusual for most data sets. Z-scores have no units.

Z-scores are found by the following formulas.

$$\text{For a sample: } z = \frac{x - \bar{x}}{s} \qquad \text{For a population: } z = \frac{x - \mu}{\sigma}$$

Example: A scientist took a sample of tree heights in a forest. His results were $s = 1.8$ meters and $\bar{x} = 9.1$ meters. Calculate the z-score for a tree that is 14 meters tall, and explain what the value means.

Solution: $z = \dfrac{14 - 9.1}{1.8} = 2.72$. This means that the tree is 2.72 standard deviations above average, it falls into the category of unusually tall (between -2 and 2), but not extreme (more than 3).

Example: Standardized IQ test scores, for the overall population, have $\sigma = 15$ points and $\mu = 100$ points. Andy Warhol was a famous artist and leader of the 70's pop art movement. It is reported that his IQ was only 86. Calculate his z-score and explain what the value means.

Solution: $z = \dfrac{86 - 100}{15} = -0.93$. This means that his IQ is about 1 standard deviation below average, it falls into the category of usual intelligence (between -2 and 2).

In the above examples, the z-scores were rounded to two decimal places. This does not follow the round-off rule used previously. Z-scores are almost always shown to two decimal places (hundredths), because there is a special use for z-scores that require looking them up in a table. That table uses two decimal places. This table and its uses will be in a later chapter, but for now, let's get used to z-scores having their own rule of two decimal places.

Try this on your own: Calculate the z-score of a woman who is 5 feet tall if the mean height is 65 inches and standard deviation is 3 inches. Is she unusually short or not? Round Z to two decimal places.

Many measurements, including large standardized test scores and children's height and weight, report how a data value ranks among the data set. Most are from a broad series of rankings called **Percentiles**. Percentiles are the values of increasing size, that divide the data into 100 parts, each with about the same number of values.

There are 99 percentiles: $P_1, P_2, P_3, ..., P_{98}, P_{99}$. P_1 is the low value that has only one percent of the data at or below it, and 99% (almost all) of the data above it. P_{80} is the high value that has 80% (most) of the data at or below it, and only 20% of the data above it.

There are three special percentiles, called **Quartiles** (every 25th percentile), which divide the data into four equal parts. There are 3 quartiles Q_1, Q_2, Q_3. The median is the same as the second quartile Q_2, which is also the 50th percentile P_{50}.

We calculate the median as described previously. Place the data is increasing size order, and find the value in the middle. Once you find the median, then the data set will be split into two halves (lower and upper). Q_1 is the median of the first half of the data. Q_3 is the median of the second half of the data.

To find other percentiles, there are a few accepted methods. They sometimes give slightly different values, but the concept is still the same, percentiles give a good estimate of how high a value is, to be equal to or above a certain percent of all of the data. The method which I present here is one that is simple and used in many other books.

First we need to put the data set in increasing order, then find the location (position) of the particular percentile we wish to find. Once we determine the position, we take the data value in that position as the percentile. The location is calculated by the formula $L = n\left(\frac{p}{100}\right)$, where n is the size of the data set and p is the percentile value we are looking for (20, 60, 75, etc.). If L is a whole number, then the percentile will be the average of the values in positions L and $L+1$. If L is a fraction, bump L up to the next whole number and the percentile is the data value in that position.

Example: Find the 20th and 93rd percentile of the following data set:

$$45, 84, 61, 34, 94, 5, 97, 42, 34, 15, 18, 71, 8, 65, 22, 10, 71, 80$$

Solution: The data in order are: $5, 8, 10, 15, 18, 22, 34, 34, 42, 45, 61, 65, 71, 71, 80, 84, 94, 97$
There are 18 values. The location of the 20th percentile is $L = 18\left(\frac{20}{100}\right) = 3.6$, so we look in the 4th position. The 4th value in order is the data value 15, so $P_{20} = 15$. There are actually 22% of the values at or below the 4th position, but this is just an estimate for the 20th percentile. The location of the 93rd percentile is $L = 18\left(\frac{93}{100}\right) = 16.74$, so we look in the 17th position. The 17th value in order is the data value 94, so $P_{93} = 94$.

There is a special set of measurements which is used to make a graphical picture of data. It is called the **Five-Number Summary** and consists of the $Min, Q_1, Median, Q_3, Max$. The graph for this is called a **Boxplot** or Box and Whisker Diagram. The steps to create a boxplot are as follows:

1. Compute the values in the 5 number summary.

2. Draw a horizontal numberline on which the 5 numbers can be located. Use a simple scale, like 0-50 or 25-100, etc.

3. Above the numberline, mark Q_1 and Q_3 with large vertical lines above their values and connect them to make the box.

4. Mark another vertical line for the median, above its value. This splits the box into two sections.

5. Mark the Min and Max with smaller vertical lines above their values.

6. Connect the centers of the sides of the box to the Min and Max with horizontal lines(whiskers).

Example: Find the five number summary of the following data set, and sketch the boxplot.
$$45, 84, 61, 34, 94, 5, 97, 42, 34, 15, 18, 71, 8, 65, 22, 10, 71, 80$$

Solution: The data in order are: $5, 8, 10, 15, 18, 22, 34, 34, 42, 45, 61, 65, 71, 71, 80, 84, 94, 97$
There are 18 values, so the median is the average of the 9th and 10th values: $Med = \frac{42+45}{2} = 43.5$. The first quartile is the median of the first nine values, so $Q_1 = 18$ (the 5th value). The third quartile is the median of the last nine values, so $Q_3 = 71$ (the 14th value). The five number summary is: $5, 18, 43.5, 71, 97$. A good numberline could be from 0-100, with marks every 10 units. The boxplot is shown below.

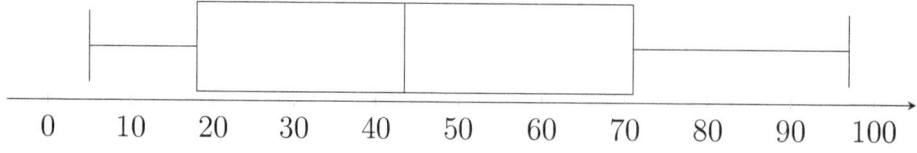

A useful interval related to the quartiles is the **Interquartile Range** (IQR), which is the range of the middle 50% of the data values, when they are in size order. $IQR = Q3 - Q1$.

Try this on your own: The grades on a science final exam were 75, 83, 96, 82, 90, 78, 60, 76, 82, 71, 92, 86, 83, 88. Calculate the 5-number summary, IQR, and sketch a regular boxplot.

Outliers were mentioned in the section on graphs. Here we will look at outliers as observations that fall well outside the overall pattern of data. Outliers are usually located far away from the mean. They may be errors in measurement, mistakenly part of the population, or just extreme values. A method for detecting outliers in a data set, is to compute upper and lower fences (described below). Any data values that lie outside the lower and upper fences, are considered to be outliers.

The **Lower Fence** (LF) of a data set is defined as $LF = Q1 - 1.5(IQR)$. It is also called the lower limit.

The **Upper Fence** (UF) of a data set is defined as $UF = Q3 + 1.5(IQR)$. It is also called the upper limit.

A **Modified Boxplot**, is a boxplot, with the addition of the fences and outliers shown as separate points. The steps to make a modified boxplot are:

1. Calculate the five number summary and the fences.

2. Determine if any data values fall outside the fences. These are outliers.

3. Determine the lowest and highest values that are NOT outliers.

4. Draw the box section of the boxplot using the quartiles.

5. Extend the whiskers out from the box, stopping at the low/high values from step 3.

6. Plot each outlier as separate points, using asterisk, dot or other marker of choice.

Example: For the following data set, find the five number summary, IQR, fences, and determine if there are any outliers. Then sketch the modified boxplot.

$$48, 99, 60, 39, 5, 124, 47, 36, 71, 29, 62, 52, 50, 42, 57$$

Solution: Data in order are: $5, 29, 36, 39, 42, 47, 48, 50, 52, 57, 60, 62, 71, 99, 124$

There are 15 values, so the median is the 8th value 50. The first quartile is the median seven values below 50, so $Q_1 = 39$. The third quartile is the median of the seven values above 50, so $Q_3 = 62$. The five number summary is: $5, 39, 50, 62, 124$.

Then $IQR = 62 - 39 = 23$, $LF = 39 - 1.5(23) = 4.5$, and $UF = 62 + 1.5(23) = 96.5$. Only the low end, 5 is close to the lower fence, but not an outlier. On the high end, 99 and 124 are beyond the upper fence, so they are outliers and will be separate points. Therefore, the upper whisker will stop at the next data value down, which is 71. A good numberline could be from 0-130, with marks every 10 units. The modified boxplot is shown below.

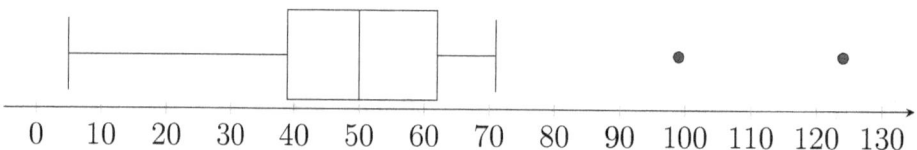

Boxplots can be used to compare two sets of data, by graphing them above one another using the same scaled axis. To tell the symmetry or shape of a data set, we generally look primarily at the box, but the whiskers can help as well.

A symmetric data set, will have a symmetric box (whiskers should be relatively symmetric). Left-skewed data will have a box that is wider on the left side. Right-skewed data will have a box that is wider on the right side. The data set that is more spread out, will have a wider overall box and the whiskers generally span out farther. Ignore outliers for comparing the spread and shape.

Example: Based on the boxplots below, state the characteristics of the data sets. Which one is more spread out?

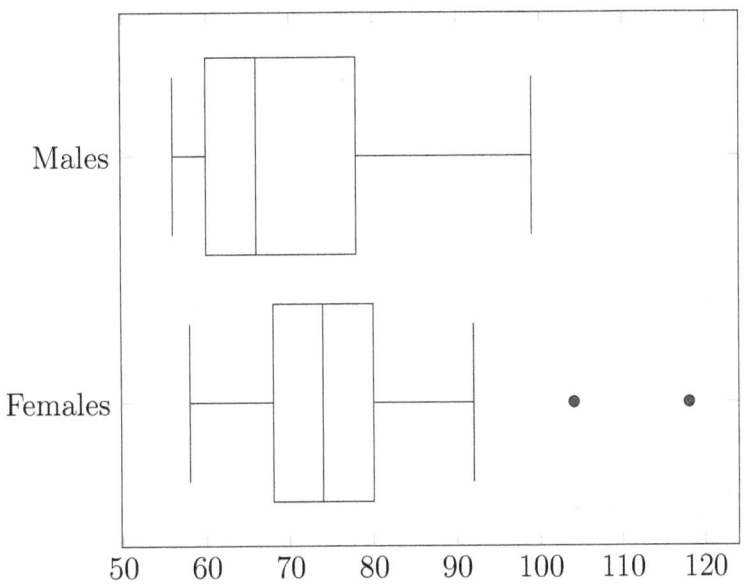

Solution: Ignoring the outliers, the female data set is symmetric and less spread out than the male data set. The male data set is right-skewed and more spread out.

Exercises: Measures of Relative Standing

Solutions appear at the end of this textbook.

1. IQ scores have a mean of 100 and a standard deviation of 15. Compute the z-scores of the famous people below and state which ones are unusual.

Name	IQ score
Garry Kasporov (chess champion)	190
Albert Einstein (scientist)	160
Arnold Schwarzenegger (actor)	135
Tim Tebow (football player)	104
Howard Stern (talk radio host)	99
George W Bush (43rd president)	125
Muhammad Ali (boxer)	78
Barack Obama (44th president)	130

2. Heights of adult males have a mean of 69 inches and a standard deviation of 3 inches. How tall must a man be to be considered unusually short or unusually tall?

3. For a data set with 38 values, compute the location (position) of the 45th percentile.

4. If a student scores at the 85th percentile on a standardized test, what does that mean?

5. For the following grades on a history test, compute the Five-Number summary.
   67, 72, 99, 100, 82, 83, 94, 90, 80, 85, 85, 77, 48, 88, 75, 50, 75, 82, and 95.

6. Create a regular box-plot for the data in the previous exercise.

7. Compute the lower and upper fences for the history test data and state which values, if any, are outliers. Then create a modified boxplot.

8. Based on the boxplots below, state the characteristics of the data sets. Which one is more spread out?

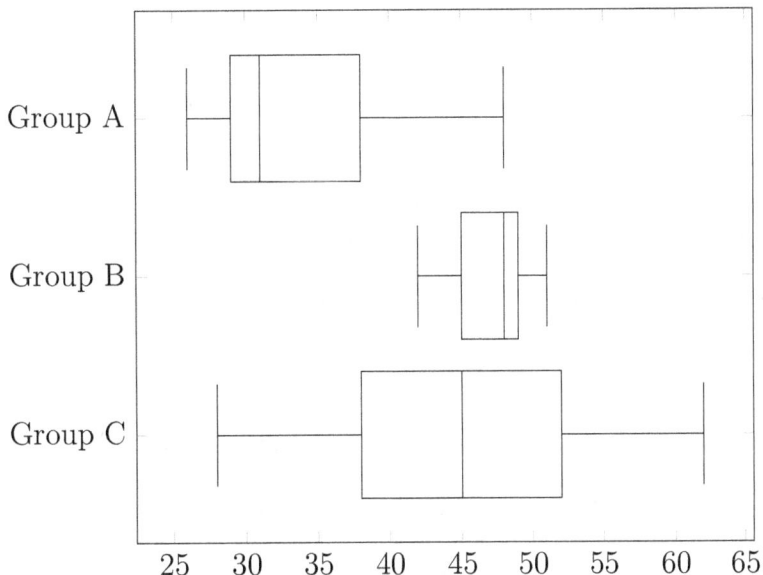

1.5 Data Sets with the TI-83 and Similar Calculators

So far, we have focused on understanding the material and working out the steps ourselves. This section will introduce the various calculator functions, which can help give quick and accurate information and graphs, especially for large data sets. It is very important to remember that a calculator is just a tool. It is quicker and more accurate than we are, but it CANNOT THINK!

If the average person were given professional tools, they could not build a house, without first learning how and gaining experience. So the average student should not be doing data analysis on a calculator, without first understanding the material and working problems out themselves (except using a calculator for large number arithmetic).

In the sampling section, you were shown how to get a random number. To get a random whole number between two values, we need to select the random integer function and input the lowest and highest values we wish to have the calculator select between. Press the $\boxed{\text{MATH}}$ button, scroll right to the **PRB** menu, scroll down and select the *randInt* function and type the low/high values in parentheses, separated by a comma.

For example, to get a random whole number between 1 and 150, follow the process above and type randInt(1,150) and hit $\boxed{\text{ENTER}}$ button. Try this yourself. Keep hitting ENTER and each time you will get another random number between 1 and 150. Occasionally numbers will repeat.

In order to have the calculator compute measures or sketch graphs, the data set must be entered into the calculator lists. The data lists can be found by pressing the $\boxed{\text{STAT}}$ button and using the **EDIT** menu. The first function **Edit**, allows you to input and view data. The default setup for data lists is a set of columns labeled L_1 (for List #1), L_2 (for List #2), etc. This allows you to store more than one set of data. Each data set will have its own list.

Whenever you ask the calculator to perform a function with a list, it is strongly recommended that you specifically tell the calculator which list to work with. If you do not, the calculator will automatically do everything with L_1. You should get into the habit of always giving the calculator specific instructions. You need to be the master to make sure you are getting what you want. Do not let the calculator programming decide for you.

Before using the lists, it is a good idea to clear out any previous data, and start with blank lists. To do this, press the $\boxed{\text{STAT}}$ button and under the **EDIT** menu, scroll down to the clear list function **ClrList** and press $\boxed{\text{ENTER}}$. This will bring up the command *ClrList* on the screen. This command must be followed by list names, so the calculator knows which lists to clear out. To put the list names on the screen, you need to press the keys $\boxed{\text{2nd}}$ and $\boxed{1}$ to type L_1 or $\boxed{\text{2nd}}$ and $\boxed{2}$ to type L_2 and so forth.

Try this: press $\boxed{\text{STAT}}$ button and under the **EDIT** menu, scroll down to the clear list function **ClrList** and press $\boxed{\text{ENTER}}$, then press $\boxed{\text{2nd}}$ $\boxed{1}$ followed by comma button $\boxed{,}$ then $\boxed{\text{2nd}}$ $\boxed{2}$ $\boxed{,}$ $\boxed{\text{2nd}}$ $\boxed{3}$ $\boxed{,}$ $\boxed{\text{2nd}}$ $\boxed{4}$ $\boxed{,}$ $\boxed{\text{2nd}}$ $\boxed{5}$ $\boxed{,}$ $\boxed{\text{2nd}}$ $\boxed{6}$ then $\boxed{\text{ENTER}}$. This will clear L_1 through L_6.

Now we are ready to input data into a list. Press the $\boxed{\text{STAT}}$ button and under the **EDIT** menu, select the **Edit** function and press $\boxed{\text{ENTER}}$. Use the arrow keys to move to the first blank under L_1. Now type the following data set in the list, by typing each value, then hit enter to move down to the next blank space. Continue to type each value and hit enter. The data is: 21, 22, 8, 19, 24, 2, 47, 30, 27, 28, 31, 40.

The calculator functions can give us many of the computed measures from sections 1.3 and 1.4. The functions use data stored in a list. Press the $\boxed{\text{STAT}}$ button and scroll over to the right to go under the **CALC** menu, select the **1-Var Stats** function and press $\boxed{\text{ENTER}}$. This will bring up the command *1-Var Stats* on the screen. Now we want to tell the calculator which list we want it to do the stats for. After the *1-Var Stats* command, type the appropriate list L_1, etc. Then hit $\boxed{\text{ENTER}}$.

If you input the data into L_1 as shown in the previous paragraph, you should now have the statistics results for L_1. There are many values here, so you may have to use the arrow keys to scroll down to see them all. All of the stats here are for x. How does it know this is supposed to be x and not t for tests grades or other data? Well, it doesn't, calculators can't think! The calculator does not know what the data represents, so all lists are labelled as a generic unknown x.

The first piece of information is $\bar{x} = 24.91666667$. You should recognize this as the mean. The next stat is $\sum x = 299$, which is the sum of the data values in that list. Next is $\sum x^2 = 9113$, which is the sum of the squares of the data values. Then the calculator shows $S_x = 12.29529351$ and $\sigma_x = 11.7718473$. These are standard deviations. They are not part of the Algebra 1 curriculum. Don't worry about them, you will see them in Algebra 2.

Scrolling down you will see $n = 12$ (we input 12 values), followed by the five number summary: $minX = 2$, $Q_1 = 20$, $Med = 25.5$, $Q_3 = 30.5$, and $maxX = 47$. You should try to work out and check all of these values yourself, to make sure you really understand the concepts from this chapter.

Going forward, I suggest that you work problems out yourself first, then use the calculator to quickly check your work. The real advantage of the calculator is when you are pressed for time on a major test. You can let the calculator do its job (giving quick, accurate information) and you can spend more time thinking, understanding, analyzing, applying, and evaluating the concepts. As you advance in higher mathematics and rigorous college math, you will be expected to understand and explain the concepts and use tools quickly.

The TI-83 (and similar calculators) can show three graphs from this chapter, histograms and boxplots. To display a graph, the data must first be input into a list. To get to the statistical graph menu, press 2nd Y=, which gets us to the STAT PLOTS menu. There are several plots that can be activated separately or at the same time. Select the first item textbfPlot1 and hit ENTER . This will bring us to the controls for Plot1. Move the cursor over the **On** choice and press ENTER to make **On** highlighted.

Use arrows to move down to the **Type** and scroll right to select the desired graph type. What is confusing, is that there appears to be two rows of icons for graphs, but this is just a continuation of the same row. If you try to arrow down to the second row, the cursor will move down to the next item on the menu. Just keep scrolling to the right instead, and the cursor will automatically move to the lower group of graph icons. As you may recognize, the third icon is a histogram, the fourth is a modified boxplot (shows outliers), and the fifth is a regular boxplot (no outliers shown). Select the modified boxplot.

The three graph types mentioned above, only use one data set (one list), so make sure the item **Xlist** is set to the appropriate list (L_1, etc.) that you want to use for that particular plot. The item **Mark**, is just a preference for the outlier symbol. Once you have all of the items set for the Plot1, you press the GRAPH button to see it. The GRAPH button is on the top row of calculator buttons, just below the screen on the right.

If you do not see your graph, do not worry. Most likely this is due to the data having larger values than what the screen is set for. To have the calculator automatically resize the screen to fit a stats plot, press the ZOOM button and scroll down to the item **ZoomStat** and press ENTER .

If you still have the data set $21, 22, 8, 19, 24, 2, 47, 30, 27, 28, 31, 40$ stored in L_1, try displaying the following graphs:

For histogram, press 2nd Y= (for stat plot menu), choose Plot1, turn it On, select type icon for histogram, set Xlist to L_1, then press GRAPH . If you do not see it, go to ZOOM and select ZoomStat, hit enter. You should see something like this:

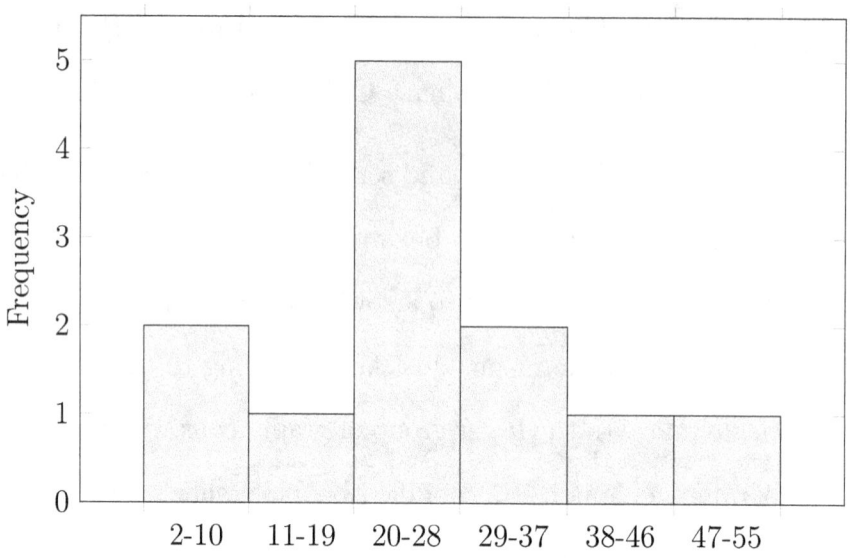

You can see graph values by pressing the TRACE button and using arrows to scroll across the graph. The values are displayed below the graph. The first bar has $min = 2$ and $max < 11$ and $n = 2$. This is the first class from 2 to 10 (less than 11) with a height of 2 (two values in this class).

The calculator sets the number of bars and the classes (grouping intervals) according to some generic programming. If you wish to have more bars (more classes) and set the width of the classes, then press the WINDOW button and change the X-axis values. Let's try Xmin=0, Xmax=56, and Xscl=7. This sets intervals of every 7 units starting at 0 and going to 56, so there will be 8 bars (some may be zero height). The new histogram should look like this:

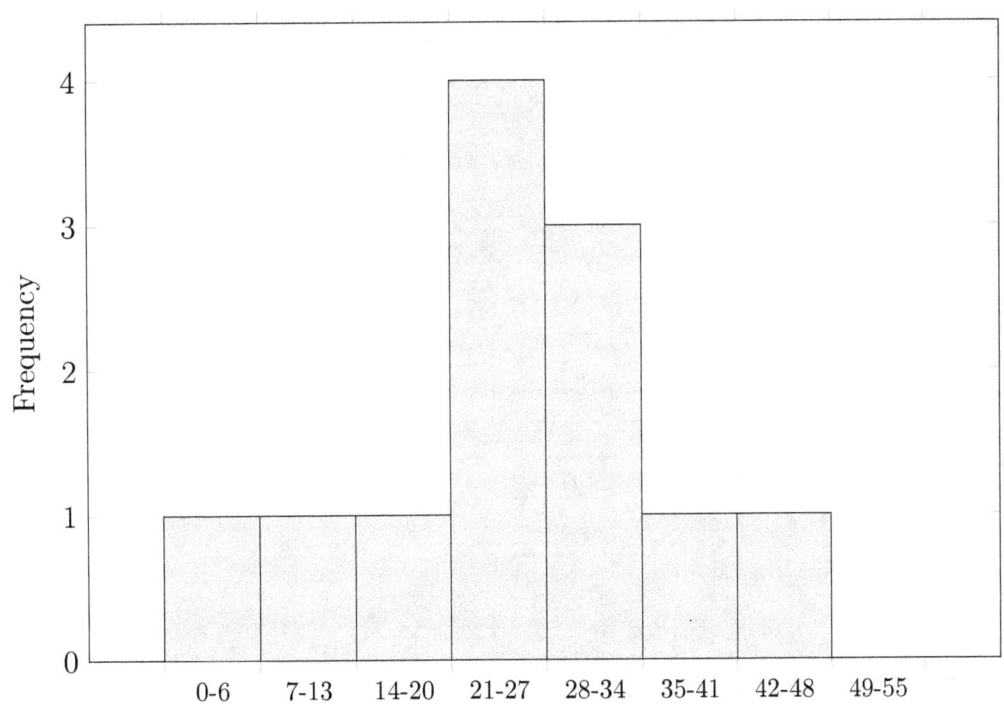

For regular boxplot, press 2nd Y= (for stat plot menu), choose Plot1, select type icon for regular boxplot (no outliers), then press GRAPH. If you do not see it, go to ZOOM and select ZoomStat, hit enter. You should see something like this:

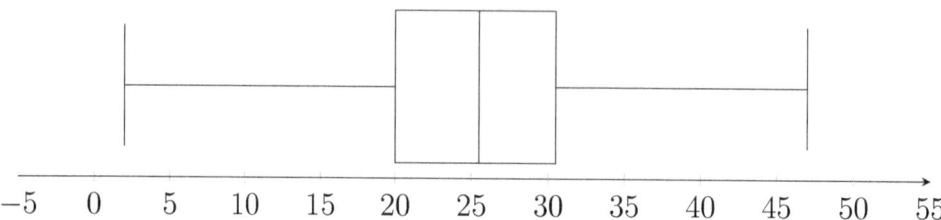

For modified boxplot, press 2nd Y= (for stat plot menu), choose Plot1, select type icon for modified boxplot (with outliers), then press GRAPH. If you do not see it, go to ZOOM and select ZoomStat, hit enter. You should see something like this:

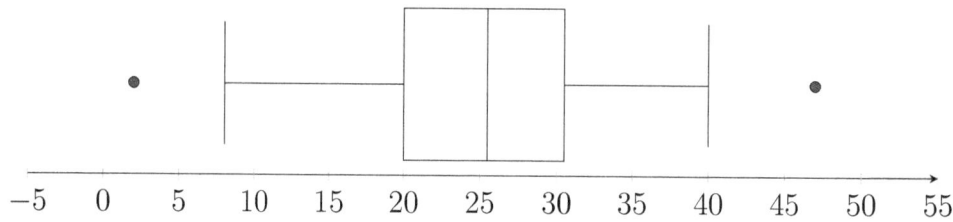

To compare two boxplots, you can turn on two plots (for example: Plot1 and Plot2) and they will graph in the same window, using the same scale. This can help determine the similarities and differences between the data sets as we saw in the previous section.

Chapter 2

Sets and Probability

2.1 Sets and Venn Diagrams

In the previous chapter, we dealt with sets of data and the science of statistics (analyzing data). Before we get into probability, we need to broaden our view of sets. What is a set? A **Set** is any collection of things. The things in the set are called **Elements**.

To state a set, we should be specific about what type of "things" we want to describe in the set, and then list all the "things" that are of that type. For example, the set of the types of items that people (in general) wear: shoes, socks, hats, shirts, pants, skirts, and so on. The set of items that YOU wear could be: glasses, Falcons jersey, Braves hat, size 34 blue jeans, etc. Another example would be types of pizza toppings: cheese, pepperoni, mushroom, chicken, etc.

There is a fairly simple mathematical notation for sets. You simply list each element, separated by a comma, and then put some curly braces around the whole set. When a set keeps going forever, or there is a long list that follows a pattern, we can use a special symbol called an **Ellipsis**. The ellipsis is a series of three dots "...", and means "continues on".

Examples: the set of natural numbers keeps going forever, and can be written as {1, 2, 3, 4, 5, ... }. The set of letters in the English alphabet can be written as {a, b, c, ..., x, y, z}. They continue, but eventually stop.

Sets, by themselves, seem pretty simple. So what do sets have to do with mathematics? Sets are the fundamental building blocks of mathematics. There are many sets of numbers, based on characteristics of those numbers. Some common mathematical sets are: even numbers $\{2, 4, 6, ...\}$, odd numbers $\{1, 3, 5, 7, ...\}$, integers $\{..., -3, -2, -1, 0, 1, 2, 3, ...\}$, prime numbers $\{2, 3, 5, 7, 11, 13, 17, ...\}$, and the list goes on.

We can write solutions as sets: positive multiples of 3 that are less than 10 $\{3, 6, 9\}$, solutions to $x^2 - 9 = 0$ are $\{-3, 3\}$, etc. There can also be sets of numbers that have no common property, they are just defined that way. For example {2, 3, 16, 828, 3839, 8827} could just be a set of random numbers.

When a mathematical set specifies a certain type of number, there has to be agreement concerning which numbers we are talking about. Mathematical sets need to be what is referred to as **Well Defined**, which means that there needs to be a straight-forward way to determine which elements are in the set and which elements are not. More general sets are not necessarily well defined.

Example: Which sets are well defined: The set of US Presidents, the set of people in charge, the set of your favorite colors?

Solution: The US president set is well defined, since there have been only certain people who have been president of the US. Even if you don't know them all, the set is known and can be looked up. The set of people in charge is not well defined, it could be just about anybody at any given time. The set of your favorite colors may be known to you right now, but other people would have no idea and it can even change for you. If it involves opinion, then a set is not well defined.

The general broad set of every element that is possible to be in a specific set, is known as the **Universal Set**. The universal set can be different, depending upon the situation. Having a universal set that is well defined, helps to easily specify particular sets. For example, if you are trying to create the set of letters in your name, we can limit the universal set to the English alphabet a, b, c, ..., x, y, z. If we did not do this, then technically, we might have to go through every alphabet in every language.

When talking about even, odd, and prime, the universal set is the set of Natural Numbers. If we are looking at a set of times for sprinters running the 100 meter dash, then the universal set could be all positive real numbers (down to 0.001 precision).

It is fairly standard to use Capital Letters to represent sets, and lowercase letters to represent an element in that set. We can use generic letters to represent some arbitrary sets like set A, set B, set C, etc. And generic lowercase letters to list the elements. Example $A = \{a, b, c, d, e\}$. Often we will try to use letters that make sense for the set. You can use C to represent the set of classes you are taking. We can use E to represent even numbers and O for the odd numbers.

Two sets are equal if they have exactly the same members. Now at first glance they may not seem equal, you may have to examine them closely! For example, are A and B equal? where A is the set whose members are the first four positive whole numbers, and $B = \{4, 2, 1, 3\}$. Let's check. They both contain 1. They both contain 2 and 3, and 4. We have checked every element of both sets, so, yes, they are equal! The equal sign is used to show equality, so you would write $A = B$. In sets, it does not matter what order the elements are in. Example: $\{1,2,3,4\}$ is the exact same set as $\{3,1,4,2\}$

When we define a set, if we take pieces of that set, we can form what is called a **Subset**. By definition, a set A is a subset of set B, if and only if every element of A is in B. So for example, we have the set $\{1, 2, 3, 4, 5\}$. A subset of this is $\{1, 2, 3\}$. Another subset is

{3, 4} or even another is simply {1}. However, {1, 6} is not a subset, since it contains an element 6 which is not in the parent set.

Example: if $A = \{1, 2, 3, 4\}$ and $B = \{1, 4, 3\}$, which set is a subset of the other (if any)?

Solution: every element of B is in A, so B is a subset of A. Set A has an extra element, so it is not a subset of B.

Example: Let A be all multiples of 4 and B be all multiples of 2. Is A a subset of B? And is B a subset of A?

Solution: Well, we can't check every element in these sets, because they have an infinite number of elements. So we need to get an idea of what the elements look like in each, and then compare them. $A = \{4, 8, 12, 16, ...\}$ and $B = \{2, 4, 6, 8, ...\}$. By pairing off members of the two sets, we can see that every member of A is also a member of B, but every member of B is not a member of A. Therefore, A is a subset of B.

Set A is a **Proper Subset** of set B, if and only if, every element in A is also in B, and there exists at least one element in B that is not in A. So a subset is any part of a set (including the entire set) where a proper subset is just part of a set (cannot be the entire set). For example {1, 2, 3} is a subset of {1, 2, 3}, but is not a proper subset of {1, 2, 3}. The set {1, 2, 3} is a proper subset of {1, 2, 3, 4}, because the element 4 is not in the first set. You should notice that if A is a proper subset of B, then it is also a subset of B.

If a set has no elements in it, it is called the **Empty Set** or (Null set). The empty set is represented by the symbol ∅ or by { } (braces with no elements inside). An example of the empty set is the set of people who have walked on Mars.

A simple analogy that should help you understand subsets it this. Imagine a bag with four coins, a penny, a nickel, a dime, and a quarter. If you reach your hand in the bag and very quickly grab some coins and pull your hand out, then the possible sets of coins in your hand are subsets of the original set of four coins in the bag. You could grab no coins, just any one of the four, some combination of two or three coins, or possibly all four. A proper subset would be any set of coins you could grab (or not grab), except for the entire set.

**Try this on your own: if $A = \{1, 2, 3, 4\}$, $B = \{3, 2, 4, 1\}$ and $C = \{1, 4, 3\}$, which sets are subsets of the others and are they proper?

There are special operations that can be done with sets. The results of these operations are called compound sets. The basic types of compound sets are shown in the following examples.

The **Complement** of a set is the set of all elements that are NOT in the set, but must be in the universal set.

If the Universal set is all natural numbers, $U = \{1, 2, 3, 4, ...\}$ and the set of even numbers is $E = \{2, 4, 6, ...\}$, then the complement of set E is all natural numbers that are not even. There is a symbol for this. The complement of E is $\overline{E} = 1, 3, 5, ...$ Notice this is also just the odd numbers. An important item to note here is that in the complement of E we did not include negative numbers, fractions, etc. Why not? After all these numbers are certainly not in the set E to begin with, but they are not in the Universal set of natural numbers we started with, so they are ignored.

Another type of compound set is called the **Intersection**. It is the set of all elements common to two or more sets. The symbol for intersection is $\cap$. Intersection is often referred to as AND.

Still another is called the **Union**. It is the set all elements that are in one or more sets of a group of sets (combined set). The symbol for union is $\cup$. Union is often referred to as OR. Think of a union as a union of two groups. When they unite, they become one larger group, containing all elements from both, with the elements listed once (even if an element was in both groups).

Example: Let the Universe $U = \{0, 1, 2, 3, 4, 5, 6, 7, 8, 9\}$, $A = \{1, 2, 4, 7\}$, and $B = \{2, 4, 6, 8\}$. List the sets $\overline{A}$, $A \cap B$, and $A \cup B$.

Solution: The complement is the set of all elements from the universe, which are not in set A, so $\overline{A} = \{0, 3, 5, 6, 8, 9\}$. The intersection is the set of elements which are in both A and B, therefore, $A \cap B = \{2, 4\}$. The union is the set of elements in either A or B, which is one large set of all of their elements combined. $A \cup B = \{1, 2, 4, 6, 7, 8\}$.

There are special diagrams that represent sets called **Venn Diagrams**, named after the British mathematician John Venn. They can be shown in many ways, but most common is to show the Universal set as a large rectangle, with specific sets shown as circles inside the rectangle. The compound sets will be the parts of the circles, or outside the circles, or where they overlap, etc. Even if there are many sets in a problem, the sets included as circles should only be the ones that are relevant to a particular compound set.

The elements of the universe are shown in the diagram, placed inside or outside the appropriate circles (or parts of circles). The corresponding area which represents the compound set is usually shaded, in order to match the set being stated.

Example: Show the Venn diagrams for the three compound sets from the previous example.

Solution: For the complement of A, we draw a rectangle with one circle that represents set A. We place the elements of A inside the circle, and all other elements outside (but in the universe rectangle). Then shade outside the circle.

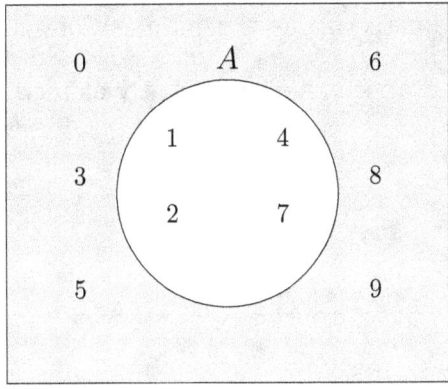

$\overline{A}$ or complement of A

For the intersection, we draw rectangle with two overlapping circles for A and B. We place the elements of A inside its circle, elements of B inside its circle, and the ones they share in the overlapping area (the intersection). Then shade the intersecting area. Notice that the shaded area has exactly two elements in it, 2 and 4, which are the stated intersection elements.

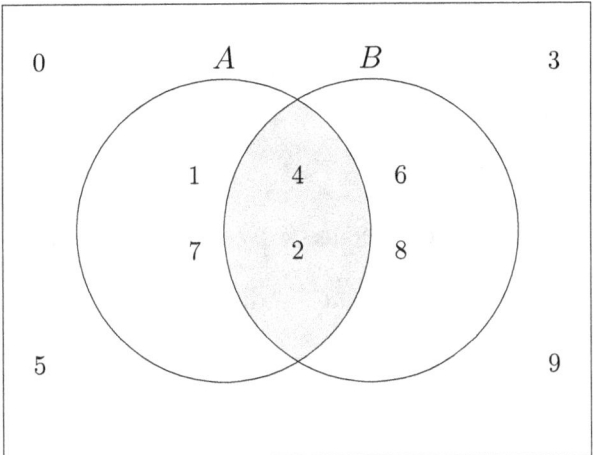

$A \cap B$ or intersection of A and B

For the union, we draw same setup as the intersection, but we shade both circles (the union). Notice that the shaded area has exactly every element from A or B in it, which are the stated union elements.

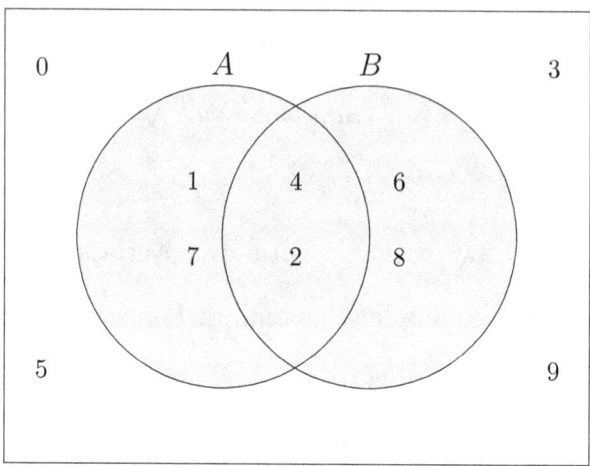

$A \bigcup B$ or union of A or B

Compound sets can get more complicated, by combining more than one symbol. The best way to handle them is to figure out each individual set that is referenced by letter and their complements (if used). Then apply the symbols using the inner symbol first to get an intermediate set, then apply the outer symbol to get the final resulting set.

Example: Let the Universe be Greek letters $U = \{\alpha, \beta, \gamma, \delta, \epsilon, \theta, \lambda, \pi, \psi, \omega\}$. $C = \{\alpha, \beta, \epsilon, \theta, \psi, \omega\}$, and $D = \{\gamma, \delta, \epsilon, \pi, \psi\}$.
List the following sets and draw their Venn diagrams. $C \bigcap \overline{D}$, and $\overline{C \bigcup D}$.

Solution: For the first compound set, C and D are already given, so we just need the individual complement of D, $\overline{D} = \{\alpha, \beta, \theta, \lambda, \omega\}$. The intersection of C and $\overline{D}$ is $C \bigcap \overline{D} = \{\alpha, \beta, \theta, \omega\}$. The Venn diagram is shown on the next page.

60

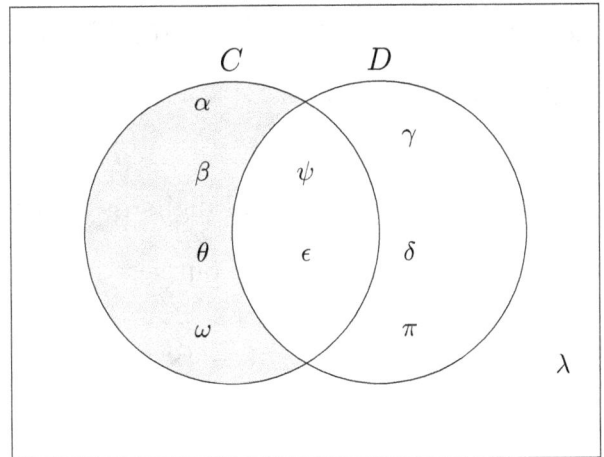

For the second compound set, we need to find the inner union first. It is the combined set of letters in either C or D, $C \bigcup D = \{\alpha, \beta, \gamma, \delta, \epsilon, \theta, \pi, \psi, \omega\}$. Now the complement of this is simply $\overline{C \bigcup D} = \{\lambda\}$. The Venn diagram is:

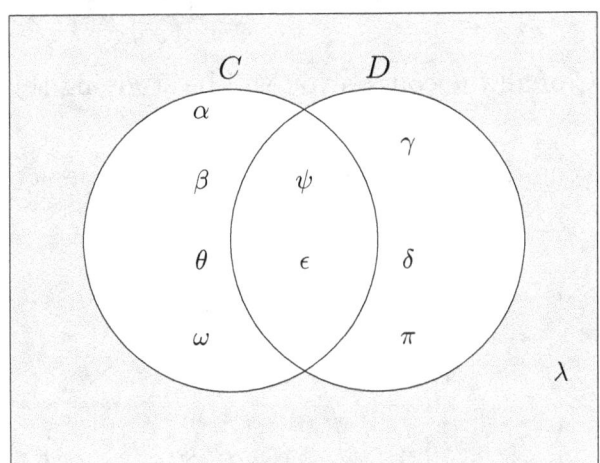

<u>Try this on your own</u>: Let the Universe $U = \{1, 2, 3, 4, 5, 6, 7, 8, 9\}$, $A = \{1, 2, 7\}$, and $B = \{2, 4, 6\}$. List the sets $\overline{B}$ and $A \bigcup B$, then sketch their Venn diagrams.

Exercises: Sets and Venn Diagrams

Solutions appear at the end of this textbook.

1. Let $U = \{0, 1, 2, 3, 4, 5, 6, 7, 8, 9\}$, the set of all digits in the decimal system. Let N be the set of digits in your phone number. Write out sets N and $\overline{N}$.

2. Let $E = \{0, 2, 4, 6, 8\}$, $O = \{1, 3, 5, 7, 9\}$, $A = \{8, 6, 4, 2, 0\}$, $B = \{1, 2, 3, 4, 5\}$, $C = \{2, 4\}$, $D = \{1, 2, 3\}$, and $G = \{5\}$. Which sets are equal (if any)? Which sets are subsets of other sets? Are they proper subsets?

3. Let $U = \{a, b, ..., y, z\}$, $V = \{a, e, i, o, u\}$, $B = \{a, b, c, d, e\}$, $R = \{j, q, x, z, u\}$. Find the elements in the following sets: $V \cap B$, $B \cap R$ and $V \cup R$

4. Sketch and shade the Venn diagrams for the compound sets from the previous exercise.

5. Write out the set C of all the courses you will be studying this academic year.

6. For the following Venn diagram, which combined set is depicted by the shading?

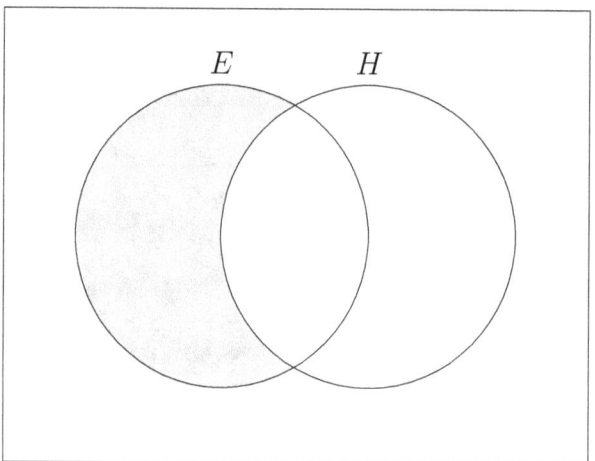

7. For the following Venn diagram, list the elements defined by the following sets?

   $\overline{A}$, $\overline{A \cap B}$, $A \cup \overline{B}$

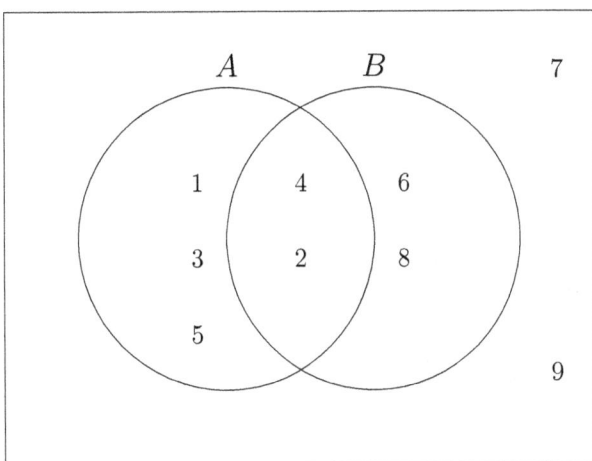

2.2 Probability Basics

In the first chapter, we learned about descriptive statistical methods for summarizing and displaying data, and descriptive measures (mean, variance, percentiles, etc.). In most studies, information about the population is the goal, but data is usually only collected from a sample, due to a census being too difficult to do. Even for single observations, we would like to know if and when something will happen. This leads us to the mathematics behind uncertainty. The science of uncertainty is called **Probability**. There are some important terms we need to know.

An **Experiment** is an action or procedure whose outcome cannot be predicted with certainty.

An **Event** is some specified result that may or may not occur when an action or procedure is performed. For example, a roll of a 6-sided die is an action. Getting a result of an even number is an event. An event is a collection of results or outcomes. In the die rolling example, the event of "even" is the collection of outcomes of the numbers 2, 4, or 6. An outcome that cannot be further broken down into components is called a **Simple Event**.

The **Sample Space** is the set of all possible outcomes of an experiment. The number of possible outcomes for a sample space is denoted by the capital letter N. All events are subsets of the sample space.

Generally we assign capital letters to represent events (A, B, C, etc.). The letter can be something meaningful such as E for even. We might assign letters in the following way. Let M = the event that someone chosen from the class is male. Let F = the event that someone chosen from the class is female. Now we can refer to M and F instead of writing out the events.

The probability value (or just probability) of an event, is the chance that the event will happen, relative to all of the possible outcomes. There are three basic properties for probability values:

1. The probability of an event is always between 0 and 1, inclusive (or 0% to 100%).

2. The probability of an event that cannot occur is equal to 0 (the event is said to be impossible).

3. The probability of an event that must occur is equal to 1 (the event is said to be certain).

When referring to the probability of an event, the capital letter P is used with the associated event (or its assigned letter) inside parentheses and placed next to the capital P. For example the probability of event male would be shown as $P(M)$.

There are three commonly used ways to calculate a probability value. The particular situation and the level of detail desired, determines which way to use.

Rule #1 for calculating a probability is the **Empirical Probability**. The probability is based on the actual results observed for some number of trials of an experiment. This is similar to relative frequency. The formula for the empirical probability of event A is $P(A) = \frac{f}{n}$, where f is the number of times the event occurred (like frequency) and n is the total number of trials of the experiment.

Example: If a women makes 14 out of 20 free-throws in her WNBA basketball tryout, what is her empirical probability of making a free-throw?

Solution: Here $n = 20$ and the number of shots she made was $f = 14$. Let F be the event of making a free-throw. Then $P(F) = \frac{14}{20} = 0.7 = 70\%$. Assuming this is what she normally does, then for the near future, her probability of making a free-throw can be estimated as 70%. If she makes the team and practices, the probability will hopefully go up.

Rule #2 for calculating a probability is the **Theoretical Probability**. This is a logical approach, but only applies to equally likely outcomes. The formula for the theoretical probability of event A is $P(A) = \dfrac{e}{N}$, where e is the number of possible outcomes that fall under the event and N is the total number of possible outcomes in the sample space.

Example: A father buys two raffle tickets for his son's baseball team raffle. There were 80 tickets sold, and one will be picked out of a hat as the winner. What is the theoretical probability of the father winning the raffle?

Solution: Here the number of possible tickets that could be picked is $N = 80$ and the number of tickets the father has is $e = 2$. Let W be the event of the father winning. Then $P(W) = \dfrac{2}{80} = 0.025 = 2.5\%$. Notice this probability was calculated BEFORE the ticket was picked. Theoretical can be reasoned out based on logic, but only when outcomes are equally likely, such as picking a ticket out of a hat.

The empirical probability of the free-throws in the previous example could not be reasoned out before we had the data of the woman's attempts. Also, making or missing the free-throw are not equally likely for most people, so empirical probability was done for that and not theoretical.

Rule #3 for calculating a probability is the **Subjective Probability**. Subjective probability is estimated by using personal knowledge or experience. It is not scientific nor mathematical and rarely logical. For example, if a young teen sneaks out of his house twice, late at night, without getting caught, he might assume his chances of getting caught are very low. Another example of this is betting on your favorite team (or against the opposing team). Most people bet with emotion and not logic.

Try this on your own: In each situation below, calculate the probability, deciding whether to use empirical or theoretical probability.

1. 200 people are at a banquet and 8 people are at your table including you. What is the probability that someone at your table is chosen at random to win a prize out of the entire banquet?

2. Danny has played 20 tennis matches this season and has won 17 of them. What is the probability that he wins his next match?

Sometimes rule #1 and #2 give about the same value, especially in the long run. This is known as the **Law of Large Numbers**, which states that as an experiment is repeated again and again, the empirical (relative frequency) probability of an event TENDS to approach the theoretical probability. For example, the theoretical probability of flipping a coin and getting the result 'Tails' is $\frac{1}{2}$ or 50% (it is one out of two equal outcomes). If you flip a coin once, the outcome will either be 100% heads or 100% tails, never 50%, but as you flip a coin many times, most likely it will be close to even for number of heads and tails.

Here is an experiment for you to try, that will demonstrate the Law of Large Numbers. Find any coin that you can distinguish one side as heads and the other as tails. Make a table on paper with 5 columns: Flips, heads, P(H), tails, P(T). For each step, fill in the values in each column, along a row and then go to the next step/row.

In step one, flip the coin five times and in the first row, put 5 under flips. Under heads, write down however many heads came up for you and same for tails. Calculate $P(H) = \frac{heads}{5}$ and $P(T) = \frac{tails}{5}$, multiplying them by 100 and rounding to nearest whole percent. For example, if you get 3 heads and 2 tails, then $P(H) = \frac{3}{5} \times 100 = 60\%$, etc. With five flips, it will be impossible to get 50%, and you could easily get values far from it.

In step two flip the coin 25 times and in the second row, put 25 under flips. Under heads, write down however many heads came up for you and same for tails. Calculate $P(H) = \frac{heads}{25}$ and $P(T) = \frac{tails}{25}$, multiplying them by 100 and rounding to nearest whole percent. For example, if you get 11 heads and 14 tails, then $P(H) = \frac{11}{25} \times 100 = 44\%$, etc. With twenty five flips, it will be impossible to get exactly 50%, but the values are most likely somewhat close. Since this is random, it could be farther away from 50% that in the first step, but unlikely.

In step three flip the coin 150 times and in the second row, put 150 under flips. Under heads, write down however many heads came up for you and same for tails. Calculate $P(H) = \frac{heads}{150}$ and $P(T) = \frac{tails}{150}$, multiplying them by 100 and rounding to nearest whole percent. For example, if you get 68 heads and 82 tails, then $P(H) = \frac{68}{150} \times 100 = 45\%$, etc. With 150 flips, it is possible to get exactly 50%, but the values are more likely to be somewhat close and probably closer that in the previous steps.

In rare cases, your values may have gotten further away instead of approaching 50%. That is why the Law states the values TEND to approach, but anything could happen. If many people did this experiment, most of them would see the values get closer to 50%. You may or may not have seen this happen. If not, try it again and I'm pretty sure it will work this time.

You may be thinking that events, outcomes, and sample spaces seem to be related to sets. Well they are actually sets. Much of what we learned in the beginning of the chapter about sets, will apply to probability and events. The Complement of an event A consists of all outcomes in which event A does NOT occur. It is denoted by $\overline{A}$. Since every outcome must either belong to set A or its complement, there is a special relationship for the probabilities. The **Complement Rule** is stated as a formula: $P(A) + P(\overline{A}) = 1$ or 100%. So once you know the probability value of one, the other is automatically determined.

Example: A bag contains 29 marbles: 3 blue, 4 red, 6 yellow, 7 orange, 4 brown, and 5 green marbles. Calculate the theoretical probability of picking a marble that is not blue. Then use the complement rule to find the same probability and compare these.

Solution: $P(\text{not blue}) = \dfrac{\#\text{not blue}}{total} = \dfrac{4+6+7+4+5}{29} = 0.897 = 90\%$

Using complement rule $P(\overline{B}) = 1 - P(B) = 1 - \frac{3}{29} = \frac{26}{29} = 0.897 = 90\%$, which is the same value as it should be, since this is the same event.

In everyday life, most people do not fully understand probability and often get it confused with a related idea called Odds. **Odds** are not direct probabilities themselves. They are the ratio of the probability that an event occurs to the probability that the event does not occur. So odds can be (and often are) greater than one. Odds are usually written as a ratio of two whole numbers and not shown as a single number, a decimal, and never a percent. Percent makes no sense for odds. Odds are commonly shown a a ratio with a colon between the values, and the values as whole numbers in lowest ratio.

The formulas for odds are:

Odds in favor of event A $= \dfrac{P(A)}{P(\overline{A})}$ Odds Against event A $= \dfrac{P(\overline{A})}{P(A)}$

Odds can be shown is three forms. During the computation the best form is a fraction $\dfrac{A}{B}$, which should be reduced so that A and B are whole numbers with no common factors. If B reduces to 1, DO NOT leave it out. Odds of $\dfrac{3}{1}$ should NEVER be shown simply as 3. Remember that odds are a relationship between to probabilities, so should always have two numbers. The other two forms are as a relation with the word "to", written as A to B, or with a colon ":", written as $A:B$. Both of these show A and B as reduced whole numbers with no common factors.

For example odds against of $\dfrac{8}{2}$ would be reduced to $\dfrac{4}{1}$ or 4 to 1 or $4:1$.

Example: Calculate the odds against, and the odds in favor of, picking a blue marble from the marble bag in the previous example.

Solution: Odds in favor of blue $= \dfrac{P(B)}{P(\overline{B})} = \dfrac{\frac{3}{29}}{\frac{26}{29}} = \dfrac{3}{26}$ or 3 : 26 odds in favor. We can state this as 3 to 26 odds in favor of blue (not very good odds). This means that out of every 29 picks, only 3 are likely to be blue and 26 not blue.

Odds against blue $= \dfrac{P(\overline{B})}{P(B)} = \dfrac{\frac{26}{29}}{\frac{3}{29}} = \dfrac{26}{3}$ or 26 : 3 odds against. We can state this as 26 to 3 odds against blue. Notice that odds against is just the reciprocal of the odds in favor (flip the fraction over). This makes it easy to calculate odds once you have one of them.

Try this on your own: If a team has a 65% chance of winning, find the odds for and against winning.

Exercises: Probability Basics

Solutions appear at the end of this textbook.

1. List the sample spaces for the following experiments:

    (a) Flipping a coin once

    (b) Flipping a coin 3 times

    (c) Rolling a 6-sided die once

    (d) Rolling two 6-sided dice and computing the sum of the dice

    (e) Randomly picking a color of the rainbow and a season of the year

2. Which of the following are valid values for a probability?

    $$0.35,\ 0.004,\ 1.23,\ 213\%,\ -0.25,\ \tfrac{3}{8},\ \tfrac{8}{3}$$

3. A student is about to roll a 6-sided die. Find the following theoretical probabilities. P(even), P(3), P(>2).

4. John is trying out for the basketball team. He shoots from the foul line 12 times. His results are: make, miss, miss, make, miss, make, miss, miss, miss, make, make, miss. What is the empirical probability of John making a shot from the foul line? What is your subjective probability of John making it onto the team?

5. If the probabilities for the types of precipitation are as follows, what is the probability of no precipitation? P(rain) = 0.20, P(snow) = 0.10, P(sleet) = 0.15.

6. If the probability of rain today is 20%, find the odds in favor of and against rain.

7. Find a 6-sided die. Roll it 15 times and compute the empirical probabilities of rolling each number (1 to 6). Then compute the theoretical probabilities of rolling each number. Compare the empirical to the theoretical. How different are they? Explain why or not.

2.3 Counting Rules

You have been counting since you were a toddler. It is pretty easy to do, or so you thought! Counting can get very complicated, especially if you had to count how many different combinations of clothing you could put together from 12 shirts, 8 pants, 3 belts, and 6 pair of shoes. Believe it or not, there are actually 1,728 different combinations of clothing (choosing one of each type). The long way to reach that value is to count each combination one at a time. The short way is to use some really cool math rules for counting.

The first one is called the **Fundamental Counting Principle**. It states that when picking one item each, from several groups of items, the total number of combinations of those items is equal to the product of how many items are in each type. In simple terms, multiply the number of items of the first type by the number of items of the second type, etc.

For the clothing example, the answer was calculated using the Fundamental Counting Principle, $12 \times 8 \times 3 \times 6 = 1,728$.

For any number of items selected from a total, the number of possible **Combinations** for a selection of r elements drawn from a population of N elements, is found using the formula $_NC_r = \dfrac{N!}{r!(N-r)!}$, where the exclamation symbol, "!", is the factorial symbol. To make sure you understand the factorial notation, $6! = 6 \times 5 \times 4 \times 3 \times 2 \times 1 = 720$. Factorial is a whole number multiplied by all whole numbers going down to 1.

Example: Compute $_{14}C_6$

Solution: $_{14}C_6 = \dfrac{14!}{6!(14-6)!} = \dfrac{14!}{6!(8!)} = \dfrac{14 \cdot 13 \cdot 12 \cdot 11 \cdot 10 \cdot 9 \cdot 8 \cdot 7 \cdot 6 \cdot 5 \cdot 4 \cdot 3 \cdot 2 \cdot 1}{6 \cdot 5 \cdot 4 \cdot 3 \cdot 2 \cdot 1 (8 \cdot 7 \cdot 6 \cdot 5 \cdot 4 \cdot 3 \cdot 2 \cdot 1)}$

Now notice that we can reduce and simplify this. The numbers 8 down to 1 on the top, will cancel off the 8 to 1 on the bottom. So the combinations $= \dfrac{14 \cdot 13 \cdot 12 \cdot 11 \cdot 10 \cdot 9}{6 \cdot 5 \cdot 4 \cdot 3 \cdot 2 \cdot 1}$

To even further simplify, we could cancel 10 and 12 on top, using the 4,3,5, and 2 on bottom. This leaves $\dfrac{14 \cdot 13 \cdot 11 \cdot 9}{6 \cdot 1} = \dfrac{18018}{6} = 3003$. So there are 3,003 different combinations of 6 elements chosen from just 14. I don't know about you, but this seems crazy even though I know it is the correct value. It is amazing how large combinations can get, just picking from a small amount like 14.

Example: A student is considering taking the following subjects this semester: Math, Physics, Literature, Economics, Psychology, and Music. How many different 4 course combinations can this student possibly take?

Solution: The student must pick 4 out of 6 courses.
$$_6C_4 = \frac{6!}{4!(6-4)!} = \frac{6 \cdot 5 \cdot 4 \cdot 3 \cdot 2 \cdot 1}{4 \cdot 3 \cdot 2 \cdot 1(2 \cdot 1)} = 15 \text{ possible schedules.}$$

A **Permutation** is an ordering of r elements selected from a set of N distinct elements. The elements selected will be in r positions. For example selecting first, second, third in a talent contest. The number of permutations of r objects chosen from a possible N is found using the formula $_NP_r = \frac{N!}{(N-r)!}$. Here, a different ordering of the same picks, is considered a different permutation.

Example: How many different permutations can be selected from a group of 10 paintings, if the judges must select a first, second, and third place award?

Solution: There are three awards, so $_{10}P_3 = \frac{10!}{(10-3)!} = \frac{10 \cdot 9 \cdot 8 \cdot 7 \cdot 6 \cdot 5 \cdot 4 \cdot 3 \cdot 2 \cdot 1}{7 \cdot 6 \cdot 5 \cdot 4 \cdot 3 \cdot 2 \cdot 1} = 720$ possible ordering of three paintings for the awards.

There are some special cases of the counting rules.

1. It is necessary for mathematicians to define $0! = 1$. This makes sense in the context of the next special case, or else the formula would lead to no answer.

2. There is only one way to select no items from a group of items, that is to not select at all. As formulas: $_NP_0 = 1$ and $_NC_0 = 1$. Since nothing is selected, there is no difference between combination or permutation (no ordering possible).

3. There are as many ways to select one item, as there are items, $_NP_1 = N$ and $_NC_1 = N$.

4. The number of ways to select all items from a group of items is just to take them all, so only one combination is possible $_NC_N = 1$.

5. There are many ways to create an ordering of all items, $_NP_N = N!$.

6. Selecting some items, is the same as leaving behind the others, so $_NC_r = {}_NC_{N-r}$. For example, selecting 5 out of 12 items is the same as leaving behind the other 7. Either way you have separated the items into two groups, one with 5 and the other 7. Notice that $_{12}C_5 = {}_{12}C_7 = 792$

Sometimes when working on word problems, it can be difficult to know if it involves combinations or permutations. The way to tell is if there is any wording that denotes some specific order of the elements. Such as award place, rank of officers (president, VP, secretary), or order based on timing (eat a donut first, then a muffin next, etc.).

If you forget which formula to use, simply calculate them both. There are always more permutations than combinations, so the larger answer comes from permutation formula. This is easy to see for a simple case of picking 3 letters out of A, B, C, D. There are only 4 combinations (leave any one of the letters out). However, there are 24 permutations, since ACB and ABC are the same combination, but different ordering.

Counting rules are useful for finding probabilities when there are a large number of outcomes. A common situation is a lottery game.

Example: There is a popular lottery ticket game called Lotto. In each play, you choose 6 different numbers from 1 to 59. To win the big jackpot, all 6 of your numbers must match the winning combination. How many different combinations of 6 numbers can be played? What is the probability of winning the jackpot with each play? What are the odds against winning?

Solution: The number of different plays of 6 numbers out of 59, is $_{59}C_6 = 45,057,474$ (over 45 million!!). Only one combination is a winner. The probability of winning $P(Win) = \frac{1}{45057474} = 0.0000000222$ (extremely small!). By the complement rule, probability of not winning the jackpot is 0.9999998778 (very large!). The odds against winning are the ratio $\frac{\frac{45057473}{45057474}}{\frac{1}{45057474}} = 45,057,473$ to 1.

NOTE: most lottery rules will incorrectly use the word odds when they are stating the probability and vice-versa. Since the probabilities are so small, there is not much difference, and most people don't understand the difference, so they don't bother to be mathematically correct.

Try this on your own: For a lottery in which you pick five numbers from 1 to 50, how many different sets can you pick if they can be in any order, and if they must be in a specific order?

Exercises: Counting Rules

Solutions appear at the end of this textbook.

1. How many different combo meals can you buy, if you get to pick one of five entrees, one of four sides, and one of three desserts?

2. Compute the following: $_8C_3$, $_{11}C_9$, $_7P_4$, $_8P_8$, $_5C_1$

3. How many different movie sequences can three friends watch, if they have 10 movies and only enough time to watch 3 of them?

4. The Fantasy 5 Lottery game consists of picking 5 different numbers from 1 to 39. How many sets of 5 numbers can be played?

5. What is the probability of winning the Fantasy 5 lottery jackpot (matching all 5 numbers), playing only once? What are the odds against winning?

6. Why is it impossible to compute $_5C_9$?

7. If a lottery game awards a $100 prize for matching any 3 out of 5 numbers, chosen from 1 to 29, what is the probability of winning?

2.4 More Probability

What chance did you think you had of being done with probability? Well sorry, there is plenty more. Some people actually get a PhD doctoral degree in probability theory. We won't go that far here, but we will look at the next level of concepts.

Events that cannot occur at the same time for one outcome of an experiment, are called **Disjoint Events** or **Mutually Exclusive**. For example, when rolling a 6-sided die, the events are rolling a 3 and rolling a 4 are mutually exclusive, because you cannot roll two numbers on one roll. You could roll a 3, then roll a 4 on the next roll, but they cannot both occur for the same roll.

Since an event and its complement never have any outcomes in common, it should be clear that complementary events are mutually exclusive. When you take a test, you either pass or fail, you can't do both at the same time. An example of events that can happen at the same time are passing a test and getting an A on the test. They are not the same, but they do share outcomes (scores of 90+). Actually, getting an A is a subset of passing.

Here is where the concepts from sets and Venn diagrams will come into play. Since events are sets of outcomes, we can combine events to get compound events with intersections and unions.

The Intersection of two events, is the set of outcomes that are part of the first event AND part of the second event at the same time. It is the set of outcomes they share in common. The symbol for intersection is $\cap$.

The Union of two events, is the set of outcomes that are part of the first event OR part of the second event (or both). It is the set of outcomes from both combined into one larger set. The symbol for union is $\cup$.

First let's look at probabilities of compound events, from a logical or reasoning perspective. We can find $P(A \text{ or } B)$ if we know the individual outcomes in each event (not just the probability values). We find the sum of the number of outcomes from A, and the number of outcomes from B, in such a way that every outcome is counted only once. Then divide this sum by the total number of outcomes in the sample space.

In a similar way, we can find $P(A \text{ and } B)$ by finding the number of outcomes that A and B both share in common. Then divide this sum by the total number of outcomes in the sample space.

Example: A class consists of 14 boys (8 are juniors, 6 are seniors) and 12 girls (8 are juniors, 4 are seniors). If one student is to be selected at random to come up to the board, find the following probabilities: P(boy $\cup$ junior) and P(girl $\cap$ senior).

Solution: The event boy $\cup$ junior is the same as boy OR junior. There are 26 students total. there are 14 boys and 16 juniors which equals 30??? How can that be? Remember that 8 of the juniors are also boys, so when we count the students for our compound event, we should only count the 14 boys (which include 8 juniors) and then the other 8 juniors (girls) to get 22 students who are boys or juniors. Now probability is $\frac{22}{26} = \frac{11}{13} = 0.846 = 84.6\%$.

The event girl $\cap$ senior is the same as girl AND senior. There are 26 students total. There are 4 girls who are seniors. The probability P(girl $\cap$ senior)= $\frac{4}{26} = \frac{2}{13} = 0.154 = 15.4\%$.

If we have the probabilities of each event, then we can find the probabilities of compound events using formulas.

The **Addition Rule** states $P(A \text{ or } B) = P(A) + P(B) - P(A \text{ and } B)$ or $P(A \cup B) = P(A) + P(B) - P(A \cap B)$, where $P(A \text{ or } B)$ is the probability at least one of the events occur in an outcome, and $P(A \text{ and } B)$ is the probability that both A and B occur at the same time in an outcome.

Example: If the probability of rain today is 0.7, the probability you forget your umbrella is 0.4, and the probability they both happen together is 0.3, what is the probability that it rains or you forget your umbrella?

Solution: P(rain or forget) $= P(rain) + P(forget) - P(both) = 0.7 + 0.4 - 0.3 = 0.8$ or 80%.

Try this on your own: A football team has 42 players. There are 18 players who play offense, 20 players who play defense, and 10 players who play on special teams. Six of the offensive players play both offense and special teams. Find the probability that a player is on the offense or special teams.

The addition rule requires that we know the intersection probability at the end of the formula. There is a rule for calculating the intersection probability directly, but before we work with that formula, we need to define **Conditional Probability**. The conditional probability of an event is the probability that results after another event has already happened and could affect the new event.

The logical way to compute conditional probability is to take into account what has already happened and adjust the possible outcomes accordingly. For example, your probability of passing a test depends upon certain conditions. If you studied well, the probability will likely increase. If you didn't realize there was a test and didn't study at all, then the probability will likely decrease. Another example could be, the probability of rolling a 5 on a 6-sided die is $\frac{1}{6}$, but the probability of rolling a 5 on a 6-sided die, after you know the roll is an even number, is 0, since 5 is odd.

If two events affect the occurrence of each other, they are said to be **Dependent**. If two events do not affect the occurrence of each other, they are said to be **Independent**.

When two events are dependent, their probabilities must be calculated using conditional probability. As a formula, the probability of the intersection of events A and B is given by $P(A \cap B) = P(A) \cdot P(B|A)$. This is known as the **Multiplication Rule**. The notation $B|A$ is read as "B given A". The formula states that the probability of both A and B, is equal to the probability of event A (considered to happen first) times the probability of event B, given that event A has already happened. P(B|A) is the conditional probability of event B, given A.

When two events are independent, then the condition of one happening does not matter, and the multiplication rule simplifies to $P(A \cap B) = P(A) \cdot P(B)$. Remember, this only happens for independent events.

Many probability problems deal with picking cards from a standard deck of playing cards. Here is description. A standard deck has 52 cards split into four symbols (called suits). The two red symbols are hearts and diamonds. The two black symbols are clubs and spades. Each suit has 13 cards with a rank (or value). The ranks are $2, 3, 4, 5, 6, 7, 8, 9, 10, J, Q, K, A$. 'J' stands for jack, 'Q" for queen, 'K' for king, and 'A' for ace. The jack, queen and king are called face cards, because they usually have faces of people on them. There are 4 of each rank card, one from each suit (symbol).

Example: If two cards selected at random from a standard deck of playing cards, what is the probability of picking two aces?

Solution: The probability of picking the first ace is $P(ace) = \frac{4}{52} = \frac{1}{13}$. The probability of the second ace being picked, depends upon the condition of an ace already being picked. P(2nd ace|1st ace) = $\frac{3}{51}$, since there would be 3 aces out 51 remaining cards. Therefore, $P(ace \cap ace) = \frac{1}{13} \cdot \frac{3}{51} = \frac{3}{663} = 0.005 = 0.5\%$.

Example: If two 6-sided dice are rolled, what is the probability of rolling two ones?

Solution: Here the two dice have no affect on each other, they are independent rolls. $P(one \cap one) = P(one) \cdot P(one) = \frac{1}{6} \cdot \frac{1}{6} = \frac{1}{36} = 0.028 = 2.8\%$.

In some situations, the conditional probability may be unknown and you wish to compute it. If the intersection probability is known, then we can rearrange the multiplication rule to find the probability of event B under the condition that event A had already happened, by $P(B|A) = \frac{P(A \cap B)}{P(A)}$. The formula states that the probability of event B, given that event A has already happened, is equal to the probability both events happen divided by the probability of event A.

Example: If one card is selected at random from a standard deck of playing cards, what is the probability of picking a Jack, given that the card picked is a face card?

Solution: Without any conditions being known, the probability of picking a jack would simply be $\frac{4}{52}$. Under these conditions, $P(jack|face) = \frac{P(jack \cap face)}{P(face)} = \frac{\frac{4}{52}}{\frac{12}{52}} = \frac{4}{12} = \frac{1}{3}$. Once you know it is a face card, it is more likely to be a jack, that just getting a jack out of all cards.

It is helpful to know all of the outcomes in a sample space for an experiment and their corresponding probabilities. A table which lists all outcomes and the probabilities is know as a **Probability Distribution**. There are two requirements for a valid probability distribution. Each probability must be between 0 and 1 (0% and 100%). The sum of all the probabilities must equal 1 or 100%. That way you know every outcome has been included properly.

Example: Which of the two tables are valid probability distributions (if any)? Why or why not?

Day	Monday	Tuesday	Wednesday	Thursday	Friday
Probability	0.22	0.13	0.34	0.24	0.07

Value	1	2	3	4	5	6
Probability	-0.3	0.3	0.4	0.4	0.1	0.1

Solution: The first table is a valid probability distribution, since each probability value is between 0 and 1, and they add up to 1. The second table is not. The probabilities do add up to 1, but one of them is negative.

For a quantitative variable, we can use the probability distribution to find the **Expected Value**. The expected value is like an average, weighted by the probabilities. It gives the typical value of the variable over many observations. It is used everyday in many ways in the real world: to set prices for insurance, to choose investments, and in most of the sciences. The formula for the expected value of a variable x is $E(x) = \sum x \cdot p(x)$, where $p(x)$ is the probability of a value x.

Example: A particular insurance policy has a claim distribution shown below. Find the expected value of a claim for a one year period. Then compute the price of the monthly insurance premium, if the company will charge 10% profit margin and the premium is paid over 12 months.

Claim Amount	$0	$500	$1,500	$5,000
Probability	0.75	0.15	0.07	0.03

Solution: $E(x) = 0(.75) + 500(.15) + 1500(.07) + 5000(.03) = \$330$, so the insurance company expects to pay out $330 on average each year, for every policy it sells. Some policies will pay out more (large claim of $5,000), most less ($0). In order to stay in business, the company must charge more than $330. The annual profit margin they charge will be 10%(330) = $33, so the annual premium is $363. Then the monthly premium is $\frac{\$363}{12}$ = $30.25. This is a simplified example of insurance, but the concept is the same as what insurance companies use everyday.

Try this on your own: A particular game has the prize distribution shown below. Find the expected value of a prize.

Prize Amount	$0	$25	$100	$500
Probability	0.7	0.2	0.09	0.01

Exercises: More Probability

Solutions appear at the end of this textbook.

1. Give an example of two events that are mutually exclusive.

2. At Mega University there are 32 physics majors, 49 math majors, and 112 engineering majors. Out of these, 8 are double majors in physics and math, and 14 are double majors in physics and engineering. Find the probability that one of these students selected at random is a physics or engineering major.

3. Given $P(A) = 0.5$, $P(B) = 0.7$, and $P(A\ and\ B) = 0.3$. Find $P(A\ or\ B)$.

4. Given $P(A) = 0.65$, $P(A\ or\ B) = 0.85$, and $P(A\ and\ B) = 0.25$. Find $P(B)$.

5. If two cards are picked at random from a deck of cards, what is the probability of picking two red sixes?

6. If one card is picked at random from a deck of cards, what is the probability of picking the ten of hearts, given that you know the card is red?

7. Give an example of two events that are independent.

8. Can two events be both mutually exclusive and independent? Explain.

9. Is this a valid probability distribution? Why or why not?

Meal	Pizza	Chicken	Steak	Pasta	Fish
Probability	0.48	0.12	0.12	0.20	0.05

10. Find the expected value of the cash prize for a lottery game. How much should they charge to play, if they want to make some profit?

Prize	$0	$5	$100	$2,500	$20,000	$100,000
Probability	$\frac{752{,}944}{800{,}000}$	$\frac{45{,}000}{800{,}000}$	$\frac{2{,}000}{800{,}000}$	$\frac{50}{800{,}000}$	$\frac{5}{800{,}000}$	$\frac{1}{800{,}000}$

Chapter 3

Other Topics

3.1 The Normal Distribution

In the beginning of the book, we learned that quantitative variables can be discrete or continuous. Probability distributions apply to discrete variables. There are only so many values and each one has a specific probability. Continuous variables must be dealt with differently. There is a continuous interval of infinitely many values or effectively infinite due to so many values in a small interval. This makes it almost impossible to calculate a probability for one particular value. Fortunately, there are formulas and tools which can give us information about probabilities of continuous variables.

Intervals of values for continuous variables have a probability weight (or density). Some intervals carry more weight than others, and as an interval increases, the weight increases. When the values are graphed relative to the probability density, we can see patterns or lack of a pattern. If a continuous variable has a graph that is symmetric and bell shaped, it can be described by a pattern and a formula. A variable that has this bell pattern is called a **Normal Distribution**.

The formula itself is very complicated and requires some seriously advanced math. Lucky for us, there are tables and calculator functions which can do the math for us. A normal distribution is completely determined by two parameters, the mean μ and the standard deviation σ.

Just to show you how difficult it is to work with the formula directly, the normal distribution probability density is given by the formula:

$$N(x) = \frac{e^{-\frac{1}{2}\left(\frac{x-\mu}{\sigma}\right)^2}}{\sigma\sqrt{2\pi}}$$

There are many real world phenomena that follow a normal distribution, leading to many different variations. However, if you look closely at the exponent of the numerator, you will see a familiar ratio $\frac{x-\mu}{\sigma}$, which is the standardized z-score of a variable x. This allows all of the different normal variables to be converted into one universal bell shape distribution called the **Standard Normal Distribution**.

The standard normal distribution has mean $\mu = 0$ and standard deviation $\sigma = 1$, which makes it very easy to work with. There is also a table that can be used to calculate values and probabilities. The table can be found at the end of this book. Many calculators have a function that can give values from the table as well.

What makes the standard normal distribution so nice to work with, is that probabilities directly correspond to area under the curve. The total area under the curve is equal to 1 or 100%. The standard normal distribution graph is shown on the next page. Notice that a large part of the graph in concentrated between $z = -1$ and $z = +1$, with most of the graph between $z = -2$ and $z = +2$ (usual values), and just about all of the graph between $z = -3$ and $z = +3$. The graph technically goes out forever, but it gets so low that effectively there is not much beyond ± 3.

The graph is shown here:

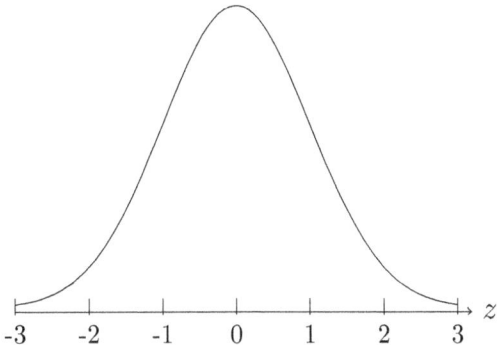

Recall from the first chapter, that a z-score is the number of standard deviations that a given value is above or below its mean. Whenever a value is below the mean, its corresponding z-score will be negative. Usual values are z-scores from -2 to $+2$. Unusual values are z-scores outside this range. Z-scores have no units. The formulas is $z = \frac{x - \mu}{\sigma}$. To match the table, z-scores are rounded to two decimal places.

When working problems applied to specific data (such as heights, IQ scores, etc.), we can convert that data into z-scores and use the standard normal distribution and table to calculate probabilities. Before we work with applied data, we need practice the procedures for the standard normal values.

The first type of problem is finding the probability corresponding to a range of z-scores. The probability is equal to the corresponding area under the bell curve, which is above the range of z-scores. The steps are listed below.

1. Draw a standard normal curve like the one shown above.

2. At the z-scores mentioned in the problem, draw vertical lines that slice the graph into sections.

3. Shade the section above the corresponding range of z-scores mentioned in the problem.

4. Look on the Standard Normal Z-table (at end of book) for the areas below (to left) of the z-scores.

On the table, the first two digits (whole and tenths) of the z-scores are listed along the left side row headings, and the hundredths are listed across the top column headings. Find the row/column that matches the z-score from the problem. Then look where that row/column meet in the body of the table to find the area (probability) to the left (below) the z-score. The areas are shown to four decimal places (the thousandths).

For example, the area to left of $z = -2.36$ is found on the first page of the table (negative z-scores), scrolling down along the left to the row for -2.3 and across under column for 0.06. The area value in this location is 0.0091. This means that the probability of a z-score being less than -2.36 is $0.0091 = 0.91\%$.

If you are looking for the probability below a z-score (to left), the answer is simply the area from table, $P(Z < \#) = area$.

If you are looking for the probability above a z-score (to right), then use complement rule, $P(Z > \#) = 1 - area$. This is the area to the right side.

If you are looking for the probability between two z-scores, then look up both table areas and subtract. $P(\#_1 < Z < \#_2) = area2 - area1$. This is the area between.

Example: Find the probability of a standard normal z-score being less than 1.5. In symbols, this is $P(Z < 1.5)$.

Solution: Draw a bell curve, make a line to slice the graph at $z = 1.5$, shade below (to left), then go to the table at end of the book. Look on the second page (positive z-scores) and go down to the row for 1.5 and across to the first column 0.00 (since the z-score is really 1.50). There we find the area of 0.9332, which is our answer, $P(Z < 1.5) = 0.9332 = 93.32\%$.

The graph is shown below. Notice that the probability is large (93.32%) and the shaded area is very large, so this makes sense.

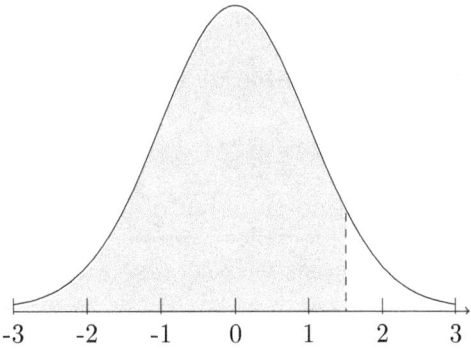

Example: Find the probability of a standard normal z-score being greater than 0.44. In symbols, this is $P(Z > 0.44)$.

Solution: Draw a bell curve, make a line to slice the graph at $z = 0.44$, shade above (to right), then go to the table at end of the book. Look on the second page (positive z-scores) and go down to the row for 0.4 and across to the fifth column 0.04. There we find the area of 0.6700, which is the area to the left, but we want the area to the right. Therefore, $P(Z > 0.44) = 1 - P(Z < 0.44) = 1 - 0.6700 = 0.3300 = 33.00\%$. The graph is shown below. Notice that the probability is somewhat small (33%) and the shaded area is somewhat small, so this makes sense.

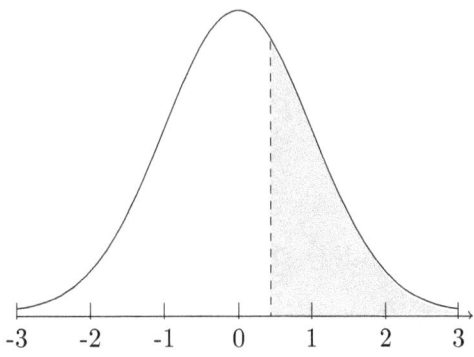

Example: Find the probability of a standard normal z-score being between -1.3 and $+1.88$. In symbols, this is $P(-1.3 < Z < 1.88)$.

Solution: Draw a bell curve, make lines to slice the graph at $z = -1.3$ and $z = 1.88$, shade above that range, then go to the table at end of the book. Look on the first page (negative z-scores) and go down to the row for -1.3 and across to the first column 0.00 (since the z-score is really -1.30). There we find the area of 0.0968, which is the area to the left of $z = -1.3$. Look on the second page (positive z-scores) and go down to the row for 1.8 and across to the column under 0.08. There we find the area of 0.9699, which is the area to the left of $z = 1.88$. Now we want the area between these, so we subtract these areas. Therefore, $P(-1.3 < Z < 1.88) = P(Z < 1.88) - P(Z < -1.30) = 0.9699 - 0.0968 = 0.8731 = 87.31\%$. The graph is shown below. Notice that the size of the probability matches the size of the shaded area, so this makes sense.

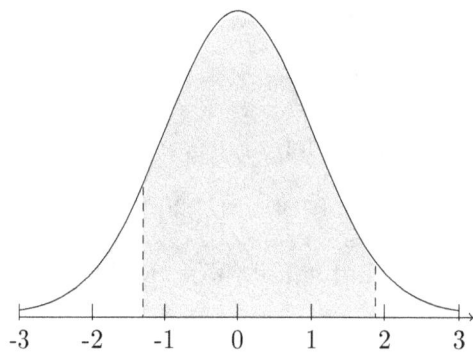

Try this on your own: Compute $P(-0.5 < Z < 1.35)$

The second type of problem is finding the z-scores which are the cutoffs for a particular area (probability). The steps are listed below. The steps are somewhat reversed from the previous type, and there are a few issues to be careful of.

1. Draw a standard normal curve (without the numbers along the axis).

2. Draw vertical lines that slice the graph into sections, to give a rough approximation of where the z-score cutoffs would have to be to match the given area or probability.

3. Shade the appropriate section of the graph (to left, to right, or between).

4. Figure out the size of other areas on the graph, so that you get a value for an area that is from a right cutoff and goes down all the way to the left end. We must do this, since the table only shows areas of this form and we need to match them.

5. Look on the Standard Normal Z-table for the left side areas we figured out in the previous step. The areas are inside the body of the table (4 decimal place numbers). Once we locate the appropriate area, then look to the edges to find the digits of the corresponding z-score. One issue that might arise, is that the table does not have every possible area value. If the area (probability) we are looking for is not there, then look for the closest area you can find, and use its z-score. If the area is exactly in the middle of two areas, then use both and average their z-scores.

Example: Find the standard normal z-score such that the probability of finding a z-score less than it, is 2.5% . In symbols, we want the z, such that $P(Z < z) = 0.025$.

Solution: Draw a bell curve, draw a line to slice the graph into a small slice way over to the left, shade below (to left), then go to the table at end of the book. Look on the first page, negative z-scores, since the cutoff is over on the negative side of the graph. Look in the body of the table for 0.0250. Notice it is in the row for -1.9 and below the column 0.06. Therefore, the z-score is -1.96. The graph is shown on the next page.

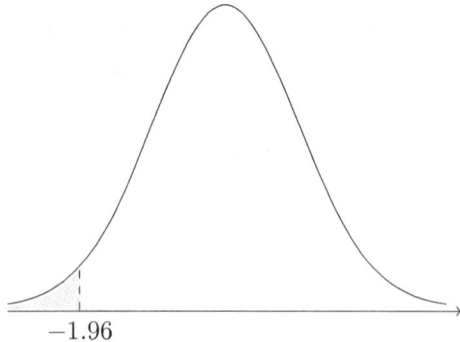

Example: Find the standard normal z-score such that the area to its right is 20%. In symbols, we want the z, such that $P(Z > z) = 0.2$.

Solution: Draw a bell curve, draw a line to slice the graph into a somewhat small slice over to the right, shade above (to right), then go to the table at end of the book. Since the table shows areas to the left (below), and our given area is to the right (above), we need to use the complement rule to convert, in order to match the table. The area to the left is $1 - 0.2 = 0.8$ (or 80%). Look on the second page, positive z-scores, since the cutoff is over on the positive side of the graph. Look in the body of the table for 0.8000. This exact value does not appear in the table, but the closest value is 0.7995. It is in the row for 0.8 and below the column 0.04. Therefore, the z-score is 0.84. The graph is shown below.

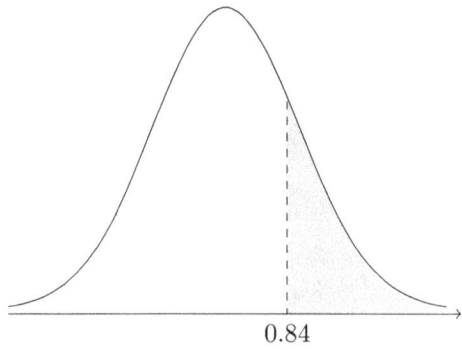

Example: Find the z-scores that cutoff the middle 90% of the graph. In symbols, we want z_1 and z_2, such that $P(z_1 < Z < z_2) = 0.9$.

Solution: Draw a bell curve, draw a line to slice the graph into mirror image small slices, one over to the right and one over to left, shade between them, then go to the table at end of the book. If 90% is between, then the other 10% is on the edges, with 5% in each tail. Then the area to the left of the low edge is 0.05. Look on the first page, negative z-scores. Look in the body of the table for 0.0500. This exact value does not appear in the table, but is exactly between 0.0505 and 0.0495. So we get both corresponding z-scores and average them. The row is -1.6 and the columns are 0.04 and 0.05. Therefore, the z-score is the average of -1.64 and -1.65. There fore $z = \frac{-1.64 + -1.65}{2} = -1.645$. The graph is shown on the next page.

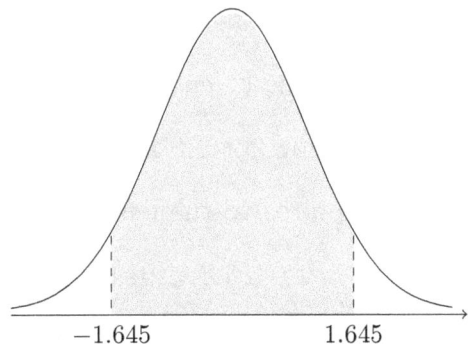

When working with actual data, such as heights or test scores, we will still use the z-table procedures, but with two important steps to add. We will use the z-score formula to convert between the data values x and the standard normal values z, as well as draw another axis below the graph to show how the x data values line up with the z-scores.

Since the z-scores are actually the number of standard deviations from the mean, then we can lineup the data mean below $z = 0$, then add/subtract the data standard deviation to put data values in line with the z-score units from -3 to 3.

Example: Adult male heights are normally distributed (follow bell curve), with a mean of $\mu = 69$ inches and a standard deviation of $\sigma = 3$ inches. What percent of men are taller than 6 feet 2 inches?

Solution: Draw a bell curve with standard z-axis from -3 to 3 and below that, an x-axis with heights that correspond to the z marks. The mean height of 69 will go below $z = 0$, one standard deviation higher (72 inches) will go below $z = 1$, two deviations higher (75) goes below $z = 2$, and 78 below $z = 3$. Do similar process on left side, subtracting standard deviation to go under the negative z-values. Change 6 feet 2 inches into 74 inches. Then convert 74 into a z-score, $z = \frac{74-69}{3} = 1.67$. Make a line to slice the graph at about $z = 1.67$, shade above (to right), then go to the table at end of the book. Look on the second page (positive z-scores) and go down to the row for 1.6 and across to the column 0.07. There we find the area of 0.9525, which is the area to the left, but we want the area to the right. Therefore, $P(X > 74) = P(Z > 1.67) = 1 - P(Z < 01.67) = 1 - 0.9525 = 0.0475 = 4.75\%$. Just less than 5% of men are taller than 6 foot 2 inches. The graph is shown below.

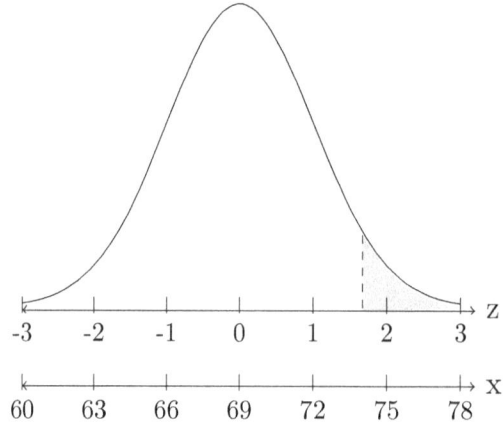

Example: The birth weights of babies in the USA are normally distributed, with a mean of $\mu = 7.4$ pounds and a standard deviation of $\sigma = 1.3$ pounds. Find the probability of a baby being born with a weight less than 5.5 pounds for a single full-term birth, which is considered to be an unhealthy birth weight.

Solution: Draw a bell curve with standard z-axis from -3 to 3 and below that, an x-axis with weights that correspond to the z marks. The mean weight of 7.4 will go below $z = 0$, one standard deviation higher (8.7 pounds) will go below $z = 1$, etc. Convert 5.5 into a z-score, $z = \frac{5.5-7.4}{1.3} = -1.46$. Make a line to slice the graph at about $z = -1.46$, shade below (to left), then go to the table at end of the book. Look on the first page (negative z-scores) and go down to the row for -1.4 and across to the column 0.06. There we find the area of 0.0721, which is the area we need. Therefore, $P(X < 5.5) = P(Z < -1.46) = 0.0721 = 7.21\%$. So 7.21% of babies in the U.S. are born too small. The graph is shown on the next page.

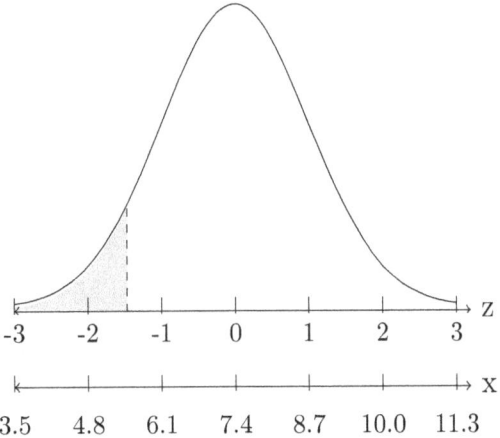

Example: IQ test scores are normally distributed, with a mean of $\mu = 100$ and a standard deviation of $\sigma = 15$. What IQ score would be at the 90th percentile?

Solution: Draw a bell curve, draw a line to slice the graph into a small slice way over to the right, shade below (to left), then go to the table at end of the book. Look on the positive z-score page, since the cutoff is over on the positive side of the graph. Look in the body of the table for 0.9000. This exact value does not appear in the table, but the closest value is 0.8997. It is in the row for 1.2 and below the column 0.08. Therefore, the z-score is 1.28. Using the formula $z = \frac{x-\mu}{\sigma}$, we can solve for the unknown IQ x. Formula setup is $1.28 = \frac{x-100}{15}$. After multiplying both sides by 15 and adding 100, we get $x = 119.2$. Rounded to whole number, the 90th percentile 119. This means that 90% of all people have an IQ lower than 119. The graph is shown below.

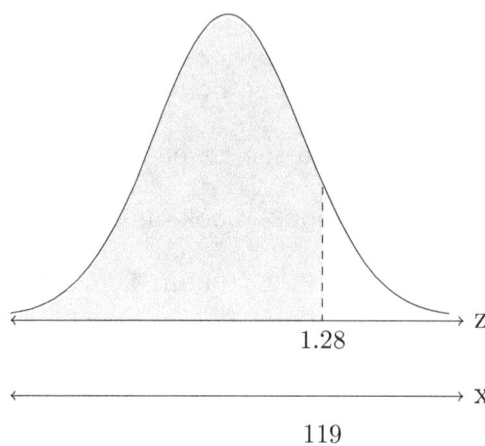

Example: A company wants to design a new specialty mountain bike. Based on research and limitations of technology, they will make a bike that is suitable for people of most every height, except for the shortest 4% of women and the tallest 1% of men. Both groups have heights that are normally distributed. Women with a mean of $\mu = 64$ inches and a standard deviation of $\sigma = 2.5$ inches. Men with a mean of $\mu = 69$ inches and a standard deviation of $\sigma = 3$ inches. What are the cutoff heights for those percentages?

Solution: Draw a bell curve, draw a line to slice the graph into small slices way over to the right and left and shade those end tails, then go to the table at end of the book. Look on the negative z-score page in the body of the table for area of 0.0400. This exact value does not appear in the table, but the closest value is 0.0401. The z-score we will use is -1.75.

For the men, 1% above the cutoff is same as 99% below. Go to positive z-score page, look in the body of the table for area of 0.9900. This exact value does not appear in the table, but the closest value is 0.9901. The z-score we will use is 2.33. Using the z-score formula for each, we get cutoff of $x = 59.6$ inches for women and 75.6 inches for men. Therefore the bike will be too big for women under 4 feet 11.6 inches and too small for men above 6 feet 3.6 inches. The combined graph is shown below.

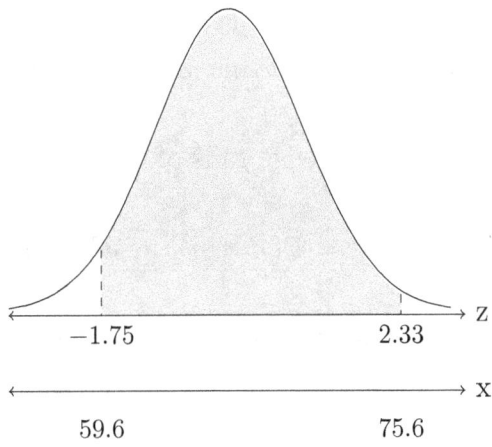

Try this on your own The birth weights of babies in Brazil are normally distributed, with a mean of $\mu = 3,110$ grams and a standard deviation of $\sigma = 463$ grams. Find the probability of a baby being born with a weight more than 3,000 grams.

Exercises: The Normal Distribution

Solutions appear at the end of this textbook.

1. Find the probability of a standard normal z-score being greater than -1.04.

2. Find $P(Z < 2.73)$.

3. Compute the area between $z = -1$ and $z = 2$.

4. Find the standard normal z-score such that the area to its right is 0.0700.

5. Find the z-scores that cutoff outer edge tails of 10% on either side of the bell curve.

6. Explain what causes $P(Z > -n) = P(Z < n)$, for all values of n.

7. Adult female heights are normally distributed, with a mean of $\mu = 64$ inches and a standard deviation of $\sigma = 2.5$ inches. What percent of women are shorter than 5 feet 6 inches?

8. The birth weights of hospital born babies in Pakistan are normally distributed, with a mean of $\mu = 2.9$ kg and a standard deviation of $\sigma = 0.5$ kg. Find the probability of a baby being born with a weight greater than 3.5 kg.

9. SAT test scores are normally distributed, with a mean of $\mu = 500$ and a standard deviation of $\sigma = 100$. What score would be at the 75th percentile?

10. Tinytown College offers three different scholarships. The silver scholarships go to applicants who score above the 85th percentile on the math portion of the SAT, but still within the 95th percentile (above that qualifies for the gold scholarship). What scores will qualify applicants for the silver scholarship?

3.2 Correlation and Regression

In previous sections, the data sets we looked at were for one variable. Many studies are done to examine the relationship between variables. Two variables measured on the same subjects are said to have a **Correlation**, when certain values of one variable tend to occur more often with certain values of the other variable. For example, people with larger heights tend to have larger weights and those with smaller heights tend to have smaller weights. This does not mean that every tall person weighs more than every short person, just that more often it happens than not. Sometimes when two variables are related, it can be that one tends to cause the other, but many times they are just related to another variable we don't know about.

Two variables are said have **Positive Correlation**, when higher values of one variable are paired with higher values of the other (and lower with lower). Two variables are said to have **Negative Correlation**, when higher values of one variable are paired with LOWER values of the other (and lower of one with HIGHER of the other).

When the variables are quantitative (numerical) we can get a picture of the relationship between them by creating a **Scatterplot**, which is a set of axes with values of one variable along the horizontal axis and values of the other along the vertical axis. Each pair of measurements for each subject are represented by a single point.

When looking at a scatterplot, we should examine the pattern of the points. We can describe the shape, strength and direction of the pattern. We should also examine the deviation from the main pattern by looking for outliers. The typical shape we look for is a line (linear relationship), where the strength would be how close to a perfect line the pattern is. Often when examining relationships, we can look at additional variables that are categorical and show them by different symbols or colors on the scatterplot. This can help distinguish different patterns for different groups, such as males and females, etc.

The chart below shows examples of strong, weak, and no linear patterns.

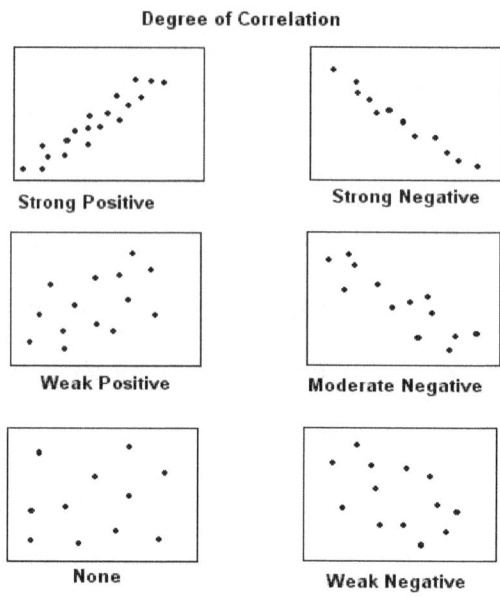

Example: The following data was collected from an actual sample of college students in a statistics class at the University of West Georgia. Create a scatterplot of the data, using different symbols for each gender (the categorical variable). What type of patterns do you see? What shape, direction and how strong are the patterns?

Gender	F	F	F	F	F	F	F	M	M	M	M	M	M
Ht (in)	59	61	62	62	65	67	70	66	67	67	71	71	73
Wt (lbs)	98	110	108	138	121	141	152	122	174	162	171	189	195

Solution: The scatterplot is below, using gray squares for the females and black triangles for the males. Overall, there is a fairly strong linear pattern in a positive direction, with no outliers. For the females, there is a strong positive linear pattern with one outlier $(62, 138)$. The male points are similar, strong positive linear pattern with one outlier $(66, 122)$.

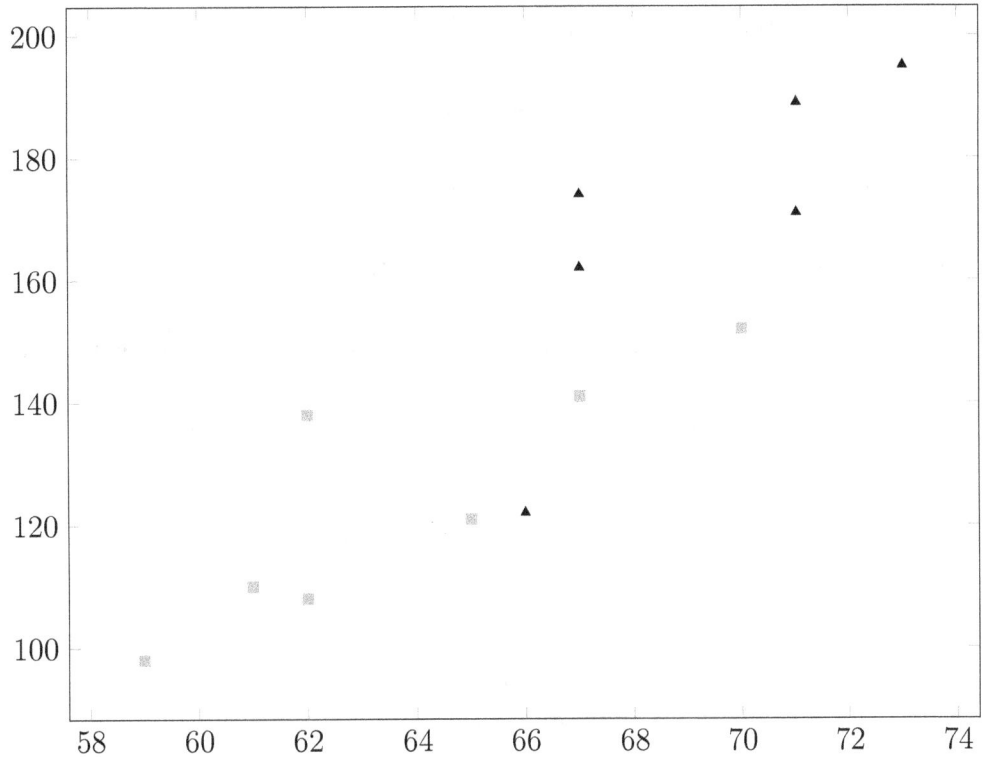

The **Linear Correlation Coefficient** measures the strength and direction of a linear relationship between two numerical variables. It is represented by the letter r.

Properties of the correlation coefficient, r:

1. It does not matter which variable is X and which is Y, the value of r is the same.

2. Changing the units of measurement does not change the value of r.

3. Positive (negative) value of r indicates a positive (negative) association.

4. r is always a value between -1 and +1.

5. A value of r close to zero indicates a weak (or no) relationship.

6. A value of r close to -1 or $+1$, indicates a strong linear relationship.

7. r is affected by outliers. Use caution if many outliers appear.

If we have two variables which we label as X and Y, there are several versions of the correlation formula. Two versions are shown here, but we will typically get r from the calculator, since even a small set of data can be time consuming and allow many chances to make mistakes. The first version of the formula uses the data values with the mean and standard deviation of each variable. The second one uses the sums of the variables and their products, which can be found from the 1-VarStats calculator function.

$$r = \frac{\sum (x - \bar{x})(y - \bar{y})}{(n-1)s_x s_y}$$

$$r = \frac{\sum xy - \frac{\sum x \sum y}{n}}{\sqrt{\left[\sum x^2 - \frac{(\sum x)^2}{n}\right]\left[\sum y^2 - \frac{(\sum y)^2}{n}\right]}}$$

So what value of r is close enough to -1 or $+1$ to say the correlation is strong? That depends upon the sample size. The table below shows the minimum absolute value of r required to say that there is a strong correlation between the variables. For the Ht/Wt data of the statistics students, $r = +0.889$, so we would be justified in saying the relationship between Ht/Wt is a strong positive one, lower heights with lower weights, and higher heights with higher weights. We can say this because the minimum value required from the table below is 0.553, and our value of 0.889 is greater than that minimum.

Critical Values for the Linear Correlation Coefficient

sample size n	3	4	5	6	7	8	9	10	11	12	13	14	15	16
minimum $|r|$	.997	.950	.878	.811	.754	.707	.666	.632	.602	.576	.553	.532	.514	.497
sample size n	17	18	19	20	21	22	23	24	25	26	27	28	29	30
minimum $|r|$	.482	.468	.456	.444	.433	.423	.413	.404	.396	.388	.381	.374	.367	.361

To show a scatterplot and the value of r on the calculator, we need to put the data into a list. Press STAT button and under EDIT menu, select the Edit function and press ENTER. Use the arrow keys to move to the first blank under L_1. Now enter the height data from the example with the stats students. Make sure you type the values in order. Then move over to L_2 and input the weight data exactly in order, so that each height matches with its correct weight. If you mixup any values, it will mess up your calculations.

Press STAT, go over to CALC menu, scroll down to item LinReg(ax+b) and press ENTER. This will put the command onto the main screen, but we also have to tell the calculator which two lists to use (or else it will give error message or even worse, it will choose for you!). Type L_1, L_2 then hit enter again. You should see the following output:

LinReg

$y = ax + b$

$a = 6.575445816$

$b = -290.8045267$

$r^2 = .7908716696$

$r = .889309659$

This is the same value $r = +0.889$, that was mentioned previously. The other values will be discussed later. If you do not see the values for r and r^2, it is just an issue of settings on the calculator. To change the settings, press 2nd 0 (for catalog menu), scroll down and select "diagnostics on", press enter until it says DONE. Now go back to STAT, CALC menu, LinReg(ax+b) and you will now see the full output.

To get the scatterplot, press 2nd Y= (for stat plot menu), choose Plot1, select first type icon for scatterplot. Set the Xlist to L_1 and Ylist to L_2, then press GRAPH. If you do not see it, go to ZOOM and select ZoomStat, hit enter. You should see a graph similar to the one shown with the ht/wt data previously in this section.

Here is an easy example of how to calculate r using the formula. The data set is small and has whole numbers. It uses the bigger version of the formula which is actually easier to work with. The terms come directly from the data and can be organized into a table of columns. The formula uses the sums of each column along with the number of data points $n = 6$.

Example: For the data below, compute the correlation r. Is the r value large enough to state that the correlation is strong?

X	Y
1	10
2	6
3	9
4	4
5	6
6	5

Solution: We fill in the values under each column, for the corresponding X and Y values, then add up each column at the bottom.

X	Y	X^2	Y^2	XY
1	10	1	100	10
2	6	4	36	12
3	9	9	81	27
4	4	16	16	16
5	6	25	36	30
6	5	36	25	30
$\sum x = 21$	$\sum y = 40$	$\sum x^2 = 91$	$\sum y^2 = 294$	$\sum xy = 125$

Now we input the value $n=6$ and all of the sums into the formula to get:

$$r = \frac{125 - \frac{21(40)}{6}}{\sqrt{\left[91 - \frac{(21)^2}{6}\right]\left[294 - \frac{(40)^2}{6}\right]}} = \frac{125 - 140}{\sqrt{[91 - 73.5][294 - 266.67]}} = \frac{-15}{21.87} = -0.686$$

We compare this to the minimum of 0.811 from the table on page 97. The value -0.686 is lower (in absolute value) than the minimum, so it is not enough to say there is a strong correlation between the variables. The correlation is a weak negative correlation.

Regression is the procedure for finding an equation (and graph) which is the best fit for a set of data. If we find that two variables have a strong linear correlation, then we would like to know the best fitting line that the scatterplot follows along. This line is called the **Regression Equation**. The formulas are very complex, so we will just use the LinReg function on the calculator to get the equation.

Fitting a line to a data set is called Linear Regression, but if the data follows another pattern, there are other types of regression to use, which give an appropriate type of an equation. Some other types of regression are quadratic, cubic, exponential, logarithmic, logistic, and sinusoidal. This book will only deal with linear regression.

We can look at a scatterplot, and if it seems to follow a linear pattern, we can sketch a reasonable line along the middle of the data to estimate where the regression line might fit. It should follow the direction of the pattern and go between the values, possibly touching some of them. There should be some values above the line and others below.

Example: Sketch a reasonable line that could fit the pattern of the scatterplot below.

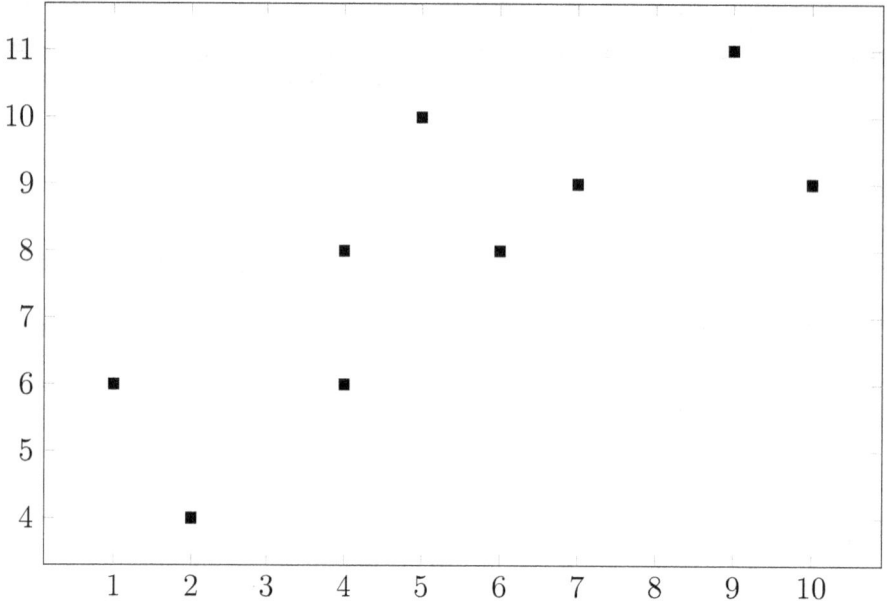

Solution: This graph shows two reasonable lines. Either one could be a good fit. Mathematically, there is a procedure to get the best fitting line, which will be introduced on the next page.

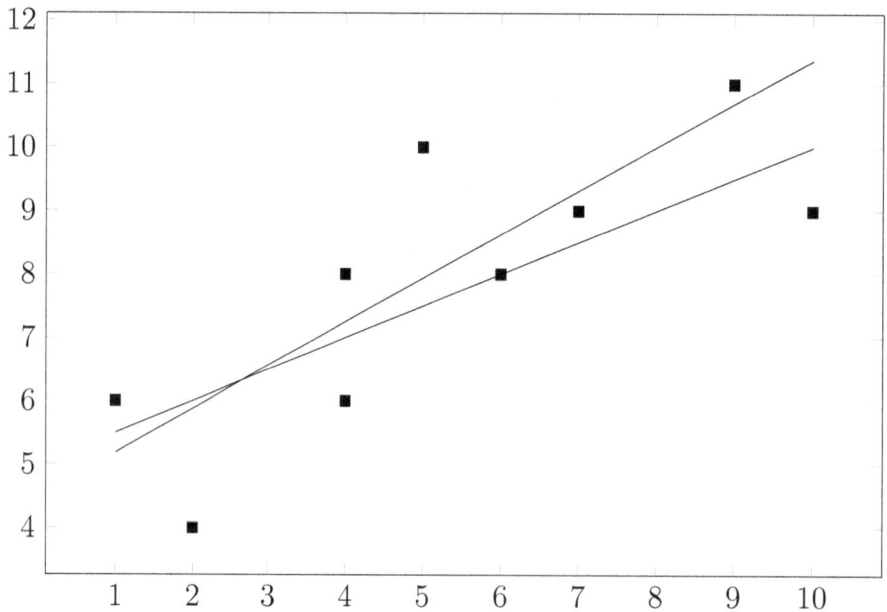

To get the best fitting linear regression equation, we use the LinReg function, which we used to get r. The equation is in the form $y = ax + b$. The value of a is the slope of the line. The value of b is the y-intercept of the line.

In the context of the data, a represents the amount of increase(+) or decrease(-) in the output y for each one unit increase of the input x, and b represents the amount of the output y when the input x is equal to zero, if the same pattern was to continue all the way down to zero. In most cases, the pattern does not continue that far, so the intercept may not be realistic, but it is needed to use the formula to compute values within the range of the data.

The regression equation can be used to predict or forecast output values for given input values. If the input value falls in the range of the original data, it is called **Interpolation**. Interpolation usually gives reasonable and realistic predictions, since the inputs are within the data range. If the input value falls outside the range of the original data, it is called **Extrapolation**. Beware of using predictions with extrapolation. For example, if you have sales data from 1999 − 2009, predicting sales for 2024 is too far into the future to be reasonable.

Example: The table below shows data for average daily sales of ice cream at a shop over a twelve month period, as well as the average number of monthly crimes in the neighborhood over the same period. Create a scatterplot and find the correlation coefficient r. Look on the table to determine if the value of r is large enough to say there is a strong linear correlation between the variables. Then find the regression equation and use it to forecast the number of crimes output for the sales values $450 and $650. Are the forecasts interpolation or extrapolation? Do you notice a pattern between sales and crime? What could explain the pattern?

Sales $	472	426	523	514	524	501	563	526	555	594	514	370
Crimes	1037	1067	1112	1180	1179	1203	1256	1268	1253	1277	1239	923

Solution: The scatterplot is below. Overall there seems to be a fairly strong linear pattern in a positive direction. This is confirmed by $r = +0.900$ which is greater than the table value 0.576.

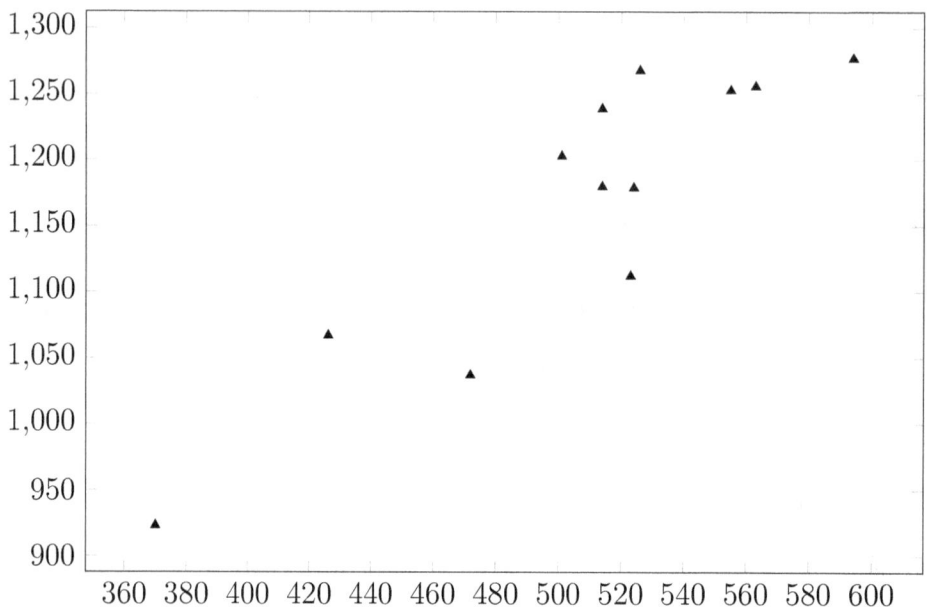

The output from the calculator shows the regression equation $y = 1.63x + 337$. This equation means that on average the number of crimes goes up by 1.63 for every extra dollar in ice cream sales, and with no ice cream sales, there would be a minimum of 337 crimes in that month.

The predictions are $y = 1.63(450) + 337 = 1071$ crimes in a month with $450 average daily sales in ice cream, and $y = 1.63(650) + 337 = 1397$ crimes in a month with $650 average daily sales in ice cream. The first one is interpolation, since 450 is in the range of the data (370 to 594). The second is extrapolation, since 650 is outside the range.

On the next page, the scatterplot is shown again, along with the best fit regression line and the two forecast values as asterisks. It is easy to graph the line, just connect the prediction points, since they came from the line equation.

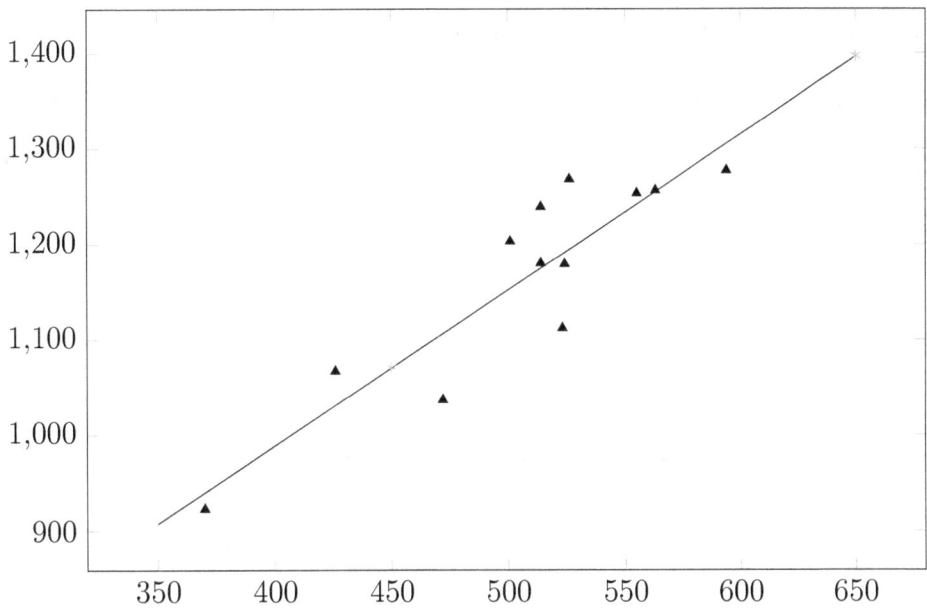

So what does all this imply about ice cream and crime? There is a strong correlation between them, so does ice cream cause more crime, or more crime make people desire ice cream? Certainly not! Remember, strong correlation does NOT mean the variables cause one another (although it is possible they could).

During the research into this data, it was found that there is actually a hidden variable that causes both. It is hot weather. When the temperature rises, people want ice cream. There are also more people on the streets and they are hot and aggravated, causing more crime. The one value that did not follow this pattern was the lowest point, which was during 100 degrees. It was too hot for most people to commit crimes or go out for ice cream.

Try this on your own: The table below shows data from ten people of their average monthly spending on fast food, as well as their average number of days of exercise each month. Create a scatterplot and find the correlation coefficient r. Then find the regression equation and use it to forecast the number of days of exercise output for the fast food value $70. Is that prediction interpolation or extrapolation? Do you notice a pattern between fast food spending and exercise? What could explain the pattern?

fast food $	20	40	58	50	140	30	90	45	100	120
exercise days	20	15	13	11	3	26	7	18	12	1

Exercises: Correlation and Regression

Solutions appear at the end of this textbook.

1. Explain what correlation is, and the difference between positive and negative correlation.

2. Match the most likely linear correlation values to the graphs below.

    $r = +0.7 \quad r = +0.99 \quad r = -0.4 \quad r = +0.15 \quad r = -0.86 \quad r = 0$

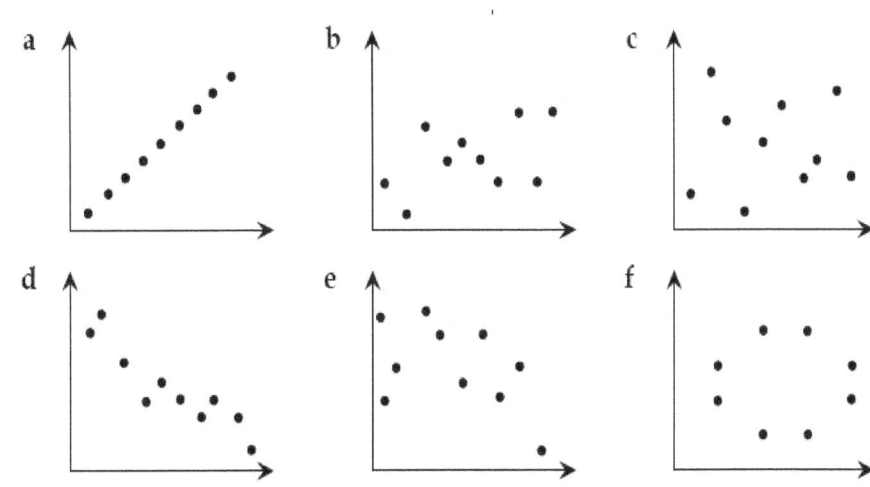

3. For the data below, use the formula to calculate the correlation r.

X	2	5	7	10	12	14
Y	2	6	7	9	11	14

4. For the data below, use the calculator to find the correlation r. Is the r value large enough to state that the correlation is strong? Create the scatterplot on the calculator.

SAT math	643	558	703	512	552	430	605
College GPA	3.52	2.91	3.63	2.21	3.02	2.80	3.18

5. Find the regression equation for the SAT/GPA data from the previous problem. Interpret the equation values. What do they say in the context of SAT scores and GPA? Add the regression line to the scatterplot on the calculator.

6. Use the regression line from the previous problem to predict GPA for math scores of 760 and 500. Are these predictions interpolation or extrapolation?

7. Draw a good estimate of the regression line for the scatterplot below. What relationship do you notice between test scores and TV watching? Which one do you think causes the other, or could there be a hidden variable causing both?

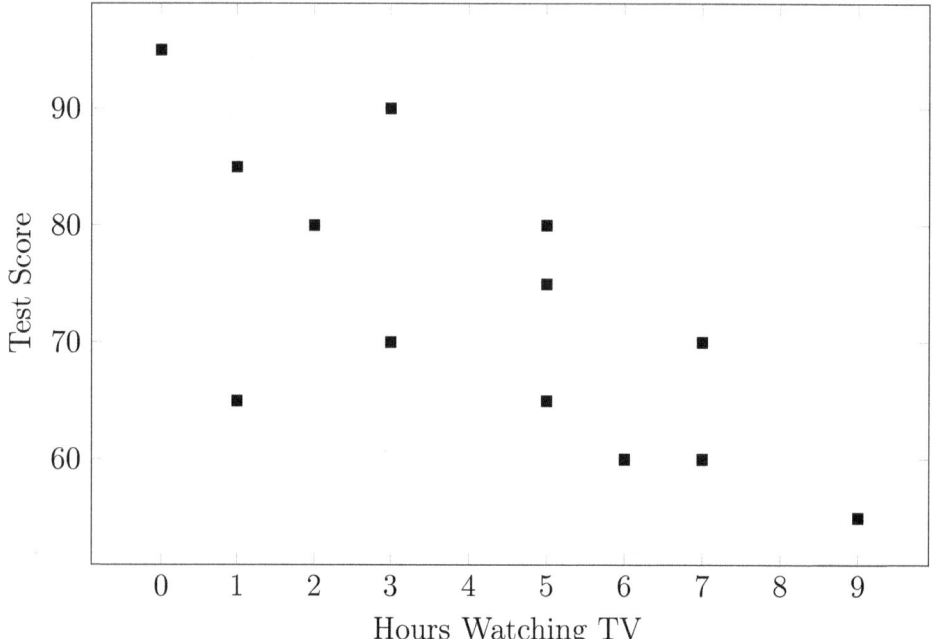

8. What is the relationship between a linear correlation coefficient r and the slope of the corresponding regression line?

3.3 Converting Units

In many of the problems in the previous sections, you had measurements that had specific units, such as a person 5 feet 6 inches tall, or a baby weighing 4 kilograms, etc. Units are very important in most fields, especially science, engineering, business, construction, transportation, and most aspects of real life. It is important to be able to understand different units and how to convert between them. Most measurements in the U.S. use what is known as the **English System of Units**.

Some of the common units used in the English System are as follows:

- For length: inch, foot, yard, mile.

- For area: acre, square inch, square yard.

- For volume: cubic foot, cup, quart, gallon, tablespoon.

- For weight: ounce, pound, ton.

The English System may be familiar to you, but in the international community, it is very confusing. It uses many different names of units and many different size ratios. Several scientists got together and came up with a simple system of units called the **Metric System**. It is used as the primary system in most countries (in the USA it is used secondary). It is based on powers of ten and uses common prefixes to represent larger and smaller variations of basic units. Once you become familiar with it, it is much easier to use and convert units in the metric system.

One of the great features of the metric system is that no matter how small or large an object is, the units used have the same base name, just with added prefixes which signify the size. On the next page is a table with the common prefixes used in the metric system.

Prefix	Symbol	Meaning
giga	G	$1,000,000,000$ times base unit
mega	M	$1,000,000$ times base unit
kilo	k	$1,000$ times base unit
hecto	h	100 times base unit
deka	da	10 times base unit
deci	d	$\frac{1}{10}$ or 0.1 of base unit
centi	c	$\frac{1}{100}$ or 0.01 of base unit
milli	m	$\frac{1}{1,000}$ or 0.001 of base unit
micro	μ	$\frac{1}{1,000,000}$ or 0.000001 of base unit
nano	n	$\frac{1}{1,000,000,000}$ or 0.000000001 of base unit

For example, the basic unit for linear measure (length) in the metric system is the **Meter**. To make larger or smaller units of length, we just need to place an appropriate prefix in front of meter. So 100 meters is also known as a hectometer. One tenth of a meter is also known as a decimeter.

To change from larger units to smaller units involves multiplying by ten repeatedly. To change from smaller units to larger units involves dividing by ten repeatedly. It also helps to have an understanding of about how big (or small) the metric units are. When doing conversions, make sure the answer makes sense. Here are some simple benchmarks to use as approximations.

- One nanometer is approximately the diameter of the largest atom (only visible under an electron microscope).

- One millimeter is approximately the width of the lead in a pencil.

- One centimeter is approximately the width of an average pinky finger.

- One meter is approximately the length of a large man's arm.

- One dekameter is approximately the length of a tractor-trailer truck (18 wheeler).

- One kilometer is approximately nine football fields end to end (including endzones).

- One Megameter is approximately the radius of the planet Earth.

When converting units, it is recommended that you setup the units as **Unit Fractions**, also known as **Conversion Ratios**. Unit fractions are ratios that show two measurements which are equal in size, but have different values with different units. For example, 1 foot is the same as 12 inches, and can be written as a unit fraction $\frac{1\,ft}{12\,in}$ or $\frac{12\,in}{1\,ft}$. Some other important lengths you should know are 1 yard = 3 feet and 1 mile = 5,280 feet.

To convert a measurement to a different unit, multiply by the appropriate unit fraction(s). The given unit of measurement that you wish to cancel, should appear in the denominator of the unit fraction, so that this unit cancels upon multiplication. The unit measurement you wish to convert into should appear in the numerator of the fraction, so that this unit will be retained upon multiplication.

Even for simple conversions, it is a good habit to use unit fractions, so you will be able to handle the more complicated conversions. The benefit of unit fractions is that it keeps everything organized and it makes it clear as to which numbers multiply and which divide. Also, if the units don't cancel properly, then you know it is setup incorrectly.

For example, it is fairly obvious that if one yard is 3 feet and each foot is 12 inches, then 1 yard is 36 inches. With unit fractions, this would look like the following:

$$1\,yard \left(\frac{3\,ft}{1\,yd}\right)\left(\frac{12\,in}{1\,ft}\right) = 1\,\cancel{yard}\left(\frac{3\,\cancel{ft}}{1\,\cancel{yd}}\right)\left(\frac{12\,in}{1\,\cancel{ft}}\right) = \frac{1(3)(12)\,in}{1(1)} = 36\,in$$

Example: Use unit fractions to convert 40,000 inches into miles.

Solution: Setup unit fractions as

$$40,000 \ in \left(\frac{1 \ ft}{12 \ in}\right)\left(\frac{1 \ mile}{5,280 \ ft}\right) = 40,000 \ \cancel{in} \left(\frac{1 \ \cancel{ft}}{12 \ \cancel{in}}\right)\left(\frac{1 \ mile}{5,280 \ \cancel{ft}}\right)$$

$$= \frac{40,000(1)(1) \ miles}{12(5,280)}$$

$$= 0.63 \ miles$$

Example: Convert 504.7 meters to kilometers and 27 meters to nanometers.

Solution: Setup unit fractions as

$$504.7 \ m \left(\frac{1 \ km}{1,000 \ m}\right) = 504.7 \ \cancel{m} \left(\frac{1 \ km}{1,000 \ \cancel{m}}\right) = \frac{504.7(1)(1) \ km}{1,000} = 0.5047 \ km$$

$$27 \ m \left(\frac{1,000,000,000 \ nm}{1 \ m}\right) = 27 \ \cancel{m} \left(\frac{1,000,000,000 \ nm}{1 \ \cancel{m}}\right)$$

$$= (27 \ nm)(1,000,000,000) = 27,000,000,000 \ nm$$

$$= 2.7 \times 10^{10} \ nm$$

Try this on your own: Convert 7 yds to inches and 310,000 cm to hectometers.

There are many approximate English to Metric Equivalents, but only one which is exact. It is 1 inch = 2.54 cm. To be accurate, we must go through this exact conversion to get the answer. If we just want an estimate, we can use one of many approximate conversions easily found in books or on the internet. Such as 1 mile = 1.6 km, etc.

Example: Convert 125 miles to kilometers, and 26,800 millimeters to inches.

Solution: For the first conversion, we start with 125 miles and multiply by unit fractions to get to feet and inches, then convert over to centimeters, then to meters and finally kilometers. Notice that the answer makes sense, because one kilometer is slightly more than half a mile, so 125 miles will be slightly less than double in kilometers.

$$125 \text{ miles} \left(\frac{5,280 \text{ ft}}{1 \text{ mile}}\right) \left(\frac{12 \text{ in}}{1 \text{ ft}}\right) \left(\frac{2.54 \text{ cm}}{1 \text{ in}}\right) \left(\frac{1 \text{ m}}{100 \text{ cm}}\right) \left(\frac{1 \text{ km}}{1,000 \text{ m}}\right)$$

$$= 125 \cancel{\text{miles}} \left(\frac{5,280 \cancel{\text{ft}}}{1 \cancel{\text{mile}}}\right) \left(\frac{12 \cancel{\text{in}}}{1 \cancel{\text{ft}}}\right) \left(\frac{2.54 \cancel{\text{cm}}}{1 \cancel{\text{in}}}\right) \left(\frac{1 \cancel{\text{m}}}{100 \cancel{\text{cm}}}\right) \left(\frac{1 \text{ km}}{1,000 \cancel{\text{m}}}\right)$$

$$= \frac{125(5,280)(12)(2.54) \text{ km}}{100(1,000)}$$

$$= 201.168 \text{ km}$$

For the second conversion, we start with millimeters and multiply by unit fractions to get to centimeters, then convert over to inches. Inches are much bigger than millimeters, so the answer should be much smaller than 26,800.

$$26,800 \text{ mm} \left(\frac{1 \text{ cm}}{10 \text{ mm}}\right) \left(\frac{1 \text{ in}}{2.54 \text{ cm}}\right)$$

$$= 26,800 \cancel{\text{mm}} \left(\frac{1 \cancel{\text{cm}}}{10 \cancel{\text{mm}}}\right) \left(\frac{1 \text{ in}}{2.54 \cancel{\text{cm}}}\right)$$

$$= \frac{26,800 \text{ in}}{10(2.54)}$$

$$= 1,055.12 \text{ in}$$

Try this on your own: Convert 8 yds to centimeters using the conversion $1 ft = 30.5 cm$.

As you probably learned in Geometry, **Area** is the amount of space enclosed inside a shape along a flat surface. Areas can be measured in square units. That is, we can still use linear units (lengths) but square them to get area.

For example, one square foot can be represented as a square shape that measures one foot on each side. Then each side is also 12 inches. This means that one square foot $= 12\ in \cdot 12\ in = 144$ square inches. However, an area of one square foot, does not have to be a square shape. It could be a long skinny rectangle (5 feet by 0.2 feet), a circle, or even an odd irregular shape. Just as long as the enclosed area is the same size.

To convert areas, we can still use conversion ratios (unit fractions), but we just have to remember that when we go from regular units to squared units, we square the numbers in the ratios as well. For example, the unit fraction $\frac{1\ ft}{12\ in}$ becomes $\frac{1\ ft^2}{144\ in^2}$, after being squared.

Example: Use unit fractions to convert 1,000 square feet to square yards.

Solution: Setup unit fractions as

$$1,000\ ft^2 \left(\frac{1\ yd}{3\ ft}\right)^2 = 1,000\ ft^2 \left(\frac{1\ yd^2}{9\ ft^2}\right)$$
$$= \frac{1,000\ yd^2}{9}$$
$$= 111.1\ yd^2$$

Example: Use unit fractions to convert 7 square meters to square centimeters.

Solution: Setup unit fractions as

$$7\ m^2 \left(\frac{100\ cm}{1\ m}\right)^2 = 7\ m^2 \left(\frac{10,000\ cm^2}{1\ m^2}\right)$$
$$= \frac{7(10,000)\ cm^2}{1}$$
$$= 70,000\ cm^2$$

When converting between the metric and English systems for area, make sure you setup unit fractions and don't forget the square them (including the numbers) to end up with the appropriate units.

Example: Convert 13 square miles to square meters.

Solution: We start with 13 square miles and multiply by unit fractions to get to feet and inches, then convert over to centimeters, then to meters.

$$13\ miles^2 \left(\frac{5,280\ ft}{1\ mile}\right)^2 \left(\frac{12\ in}{1\ ft}\right)^2 \left(\frac{2.54\ cm}{1\ in}\right)^2 \left(\frac{1\ m}{100\ cm}\right)^2$$

$$= 13\ miles^2 \left(\frac{27,878,400\ ft^2}{1\ mile^2}\right) \left(\frac{144\ in^2}{1\ ft^2}\right) \left(\frac{6.4516\ cm^2}{1\ in^2}\right) \left(\frac{1\ m^2}{10,000\ cm^2}\right)$$

$$= 13\ \cancel{miles^2} \left(\frac{27,878,400\ \cancel{ft^2}}{1\ \cancel{mile^2}}\right) \left(\frac{144\ \cancel{in^2}}{1\ \cancel{ft^2}}\right) \left(\frac{6.4516\ \cancel{cm^2}}{1\ \cancel{in^2}}\right) \left(\frac{1\ m^2}{10,000\ \cancel{cm^2}}\right)$$

$$= \frac{13(27,878,400)(144)(6.4516)\ m^2}{10,000}$$

$$= 33,669,845\ m^2$$

Both systems have special units for land areas. The English system uses a unit called an **Acre**, and 1 acre = $43,560$ square feet. The metric system uses a unit called an **Are** (pronounced like the word "air"), and 1 are = $100\ m^2$. Since this is too small for land, they typically use a hectare which is 100 are = $10,000\ m^2 = 0.01\ km^2$.

Example: The University of West Georgia campus is 644 acres. How many square feet is the campus? How many hectares?

Solution: The first one is a straight conversion from acres to square feet. Notice the unit fraction does not have to be squared, since it already has area units. Acres are large, so the answer is a large number of square feet.

$$644 \; acres \left(\frac{43,560 \; ft^2}{1 \; acre}\right) = 644 \; \cancel{acres} \left(\frac{43,560 \; ft^2}{1 \; \cancel{acre}}\right) = 644(43,560) \; ft^2 = 28,052,640 \; ft^2$$

The second one has many conversions.

$$644 \; acres \left(\frac{43,560 \; ft^2}{1 \; acre}\right)\left(\frac{12 \; in}{1 \; ft}\right)^2 \left(\frac{2.54 \; cm}{1 \; in}\right)^2 \left(\frac{1 \; m}{100 \; cm}\right)^2 \left(\frac{1 \; are}{100 \; m^2}\right)\left(\frac{1 \; hectare}{100 \; ares}\right)$$

$$= 644 \; acres \left(\frac{43,560 \; ft^2}{1 \; acre}\right)\left(\frac{144 \; in^2}{1 \; ft^2}\right)\left(\frac{6.4516 \; cm^2}{1 \; in^2}\right)\left(\frac{1 \; m^2}{10,000 \; cm^2}\right)\left(\frac{1 \; are}{100 \; m^2}\right)\left(\frac{1 \; hectare}{100 \; ares}\right)$$

$$= 644 \; \cancel{acres} \left(\frac{43,560 \; \cancel{ft^2}}{1 \; \cancel{acre}}\right)\left(\frac{144 \; \cancel{in^2}}{1 \; \cancel{ft^2}}\right)\left(\frac{6.4516 \; \cancel{cm^2}}{1 \; \cancel{in^2}}\right)\left(\frac{1 \; \cancel{m^2}}{10,000 \; \cancel{cm^2}}\right)\left(\frac{1 \; \cancel{are}}{100 \; \cancel{m^2}}\right)\left(\frac{1 \; hectare}{100 \; \cancel{ares}}\right)$$

$$= \frac{644(43,560)(144)(6.4516) \; hectares}{10,000(100)(100)}$$

$$= 260.62 \; hectares$$

Another geometry concept you should have learned is **Volume**, which is the amount of space inside a three dimensional shape. Think of it as how much liquid it would take to fill the space. Volumes can be measured in cubic units. That is, we can still use linear units (lengths) but raise them to the third power to get volume.

For example, one cubic yard can be represented as a cube shape that measures one yard on each side. Then each side is also 3 feet. This means that one cubic yard = $3 \; ft \times 3 \; ft \times 3 \; ft =$ 27 cubic feet. To convert, we can still use conversion ratios (unit fractions). We just have to remember that when we go to cubic units, we cube the numbers as well.

Example: Use unit fractions to convert 900,000 cubic yards to cubic miles.

Solution: Setup unit fractions as

$$900,000 \ yd^3 \left(\frac{3 \ ft}{1 \ yd}\right)^3 \left(\frac{1 \ mile}{5,280 \ ft}\right)^3$$

$$= 900,000 \ \cancel{yd^3} \left(\frac{27 \ \cancel{ft^3}}{1 \ \cancel{yd^3}}\right) \left(\frac{1 \ mile^3}{147,197,952,000 \ \cancel{ft^3}}\right)$$

$$= \frac{900,000(27) \ mile^3}{147,197,952,000}$$

$$= 0.000165 \ mile^3$$

Surprising that so many cubic yards is only a small fraction of a cubic mile. This means that a cubic mile is HUGE! As a reference, it is the amount of water inside Devil's Lake in North Dakota, which is a large lake with fishing, boating, and a state park on an island in the middle of it.

Example: Convert $2 \ km^3$ to cm^3.

Solution: Setup unit fractions as

$$2 \ km^3 \left(\frac{1,000 \ m}{1 \ km}\right)^3 \left(\frac{100 \ cm}{1 \ m}\right)^3$$

$$= 2 \ \cancel{km^3} \left(\frac{1,000,000,000 \ \cancel{m^3}}{1 \ \cancel{km^3}}\right) \left(\frac{1,000,000 \ cm^3}{1 \ \cancel{m^3}}\right)$$

$$= 2(1,000,000,000)(1,000,000) cm^3$$

$$= 2 \times 10^{15} \ cm^3$$

When converting between the metric and English systems for volume in cubic units, make sure you setup unit fractions and don't forget the cube them (including the numbers) to end up with the appropriate units.

Example: Convert $7,000\ mm^3$ to in^3.

Solution: Setup unit fractions as

$$7,000\ mm^3 \left(\frac{1\ cm}{10\ mm}\right)^3 \left(\frac{1\ in}{2.54\ cm}\right)^3$$

$$= 7,000\ mm^3 \left(\frac{1\ cm^3}{1,000\ mm^3}\right)\left(\frac{1\ in^3}{16.387\ cm^3}\right)$$

$$= \frac{7,000}{1,000(16.387)} in^3$$

$$= 0.43\ in^3$$

Both systems have special units for measuring volume. The English system uses many different units. Some of the common units, their abbreviations and their equivalents are:

1 teaspoon (tsp)		
1 tablespoon (Tbsp)	= 3 teaspoons	
1 fluid ounce (fl.oz.)	= 2 Tbsp	
1 cup	= 8 fl.oz.	
1 pint	= 2 cups	= 16 fl.oz.
1 quart	= 2 pints	= 32 fl. oz.
1 gallon	= 4 quarts	= 128 fl. oz.
1 cubic foot	= 7.5 gallons	(rounded)

Example: How many cups in a gallon?

Solution: Setup unit fractions as

$$1\ gal \left(\frac{128\ fl.oz}{1\ gal}\right)\left(\frac{1\ cup}{8\ fl.oz}\right)$$

$$= 1\ \cancel{gal} \left(\frac{128\ \cancel{fl.oz}}{1\ \cancel{gal}}\right)\left(\frac{1\ cup}{8\ \cancel{fl.oz}}\right)$$

$$= \frac{1(128)}{8}\ cups$$

$$= 16\ cups$$

Example: How many quarts in 50 cubic feet?

Solution: Setup unit fractions as

$$50\ ft^3 \left(\frac{7.5\ gal}{1\ ft^3}\right)\left(\frac{4\ qts}{1\ gal}\right)$$

$$= 50\ \cancel{ft^3} \left(\frac{7.5\ \cancel{gal}}{1\ \cancel{ft^3}}\right)\left(\frac{4\ qts}{1\ \cancel{gal}}\right)$$

$$= \frac{50(7.5)(4)}{1}\ qts$$

$$= 1500\ quarts$$

The base unit of volume in the metric system is called a **Liter** (abbreviated as L). The Liter was set to be the exact volume of a cube, 10 centimeters on each side. On a smaller scale, 1 milliliter = 1 cubic centimeter, which is also known as a "cc". The same prefixes and factors that were used for Meters, are also used for Liters. This is another reason that the Metric System is easy to use.

Example: Convert 3.5 kiloliters to milliliters.

Solution: Setup unit fractions as

$$3.5 \, kL \left(\frac{1,000 \, L}{1 \, kL} \right) \left(\frac{1,000 \, mL}{1 \, L} \right)$$

$$= 3.5 \, \cancel{kL} \left(\frac{1,000 \, \cancel{L}}{1 \, \cancel{kL}} \right) \left(\frac{1,000 \, mL}{1 \, \cancel{L}} \right)$$

$$= 3.5(1,000)(1,000) \, mL$$

$$= 3,500,000 \, mL$$

We can also switch between standard volume units and cubic units in the metric system. In the next example, notice that one kiloliter (1,000 Liters) turns out to be the same as one cubic meter.

Example: Convert $4 \, m^3$ to Liters.

Solution: We convert to cubic centimeters, which are same as milliliters, then to Liters.

$$4 \, m^3 \left(\frac{100 \, cm}{1 \, m} \right)^3 \left(\frac{1 \, mL}{1 \, cm^3} \right) \left(\frac{1 \, L}{1,000 \, mL} \right)$$

$$= 4 \, \cancel{m^3} \left(\frac{1,000,000 \, \cancel{cm^3}}{1 \, \cancel{m^3}} \right) \left(\frac{1 \, \cancel{mL}}{1 \, \cancel{cm^3}} \right) \left(\frac{1 \, L}{1,000 \, \cancel{mL}} \right)$$

$$= \frac{4(1,000,000)}{1,000} L$$

$$= 4,000 \, L$$

There are some useful conversions between English and metric standard volume units. They are not exact, but are very close and good enough for most purposes. The main conversion that we will use is $1 \, L = 33.8 \, fl.oz.$

Example: How many Liters in 1 gallon?

Solution: We convert to fluid ounces, then to Liters. Notice that a Liter is turns out to be slightly more than a quart. You should be familiar with this from drinking bottled beverages.

$$1 \, gal \left(\frac{128 \, fl.oz}{1 \, gal} \right) \left(\frac{1 \, L}{33.8 \, fl.oz} \right)$$

$$= 1 \, \cancel{gal} \left(\frac{128 \, \cancel{fl.oz}}{1 \, \cancel{gal}} \right) \left(\frac{1 \, L}{33.8 \, \cancel{fl.oz}} \right)$$

$$= \frac{1(128)}{33.8} L$$

$$= 3.79 \, L$$

Example: How many teaspoons in 1 Liter?

Solution: We convert a Liter to fluid ounces, then to tablespoons, then finally teaspoons. Consider that it should take quite a few teaspoons to fill a Liter bottle.

$$1 \, L \left(\frac{33.8 \, fl.oz}{1 \, L} \right) \left(\frac{2 \, Tbsp}{1 \, fl.oz} \right) \left(\frac{3 \, tsp}{1 \, Tbsp} \right)$$

$$= 1 \, \cancel{L} \left(\frac{33.8 \, \cancel{fl.oz}}{1 \, \cancel{L}} \right) \left(\frac{2 \, \cancel{Tbsp}}{1 \, \cancel{fl.oz}} \right) \left(\frac{3 \, tsp}{1 \, \cancel{Tbsp}} \right)$$

$$= 1(33.8)(2)(3) \, tsp$$

$$= 202.8 \, tsp$$

All of the previous units and conversions dealt with the size of objects, or how much space they took up. Two other characteristics of objects are **Mass** and **Weight**. Mass is a measure of how much stuff an object is made of, and weight is the measure of how strongly gravity pulls on an object.

On the planet earth, mass and weight have a specific relationship, and so their units can be converted. However, that relationship changes in different gravity (in space, on the moon, etc.). For example, I (the author) weigh about 150 pounds on Earth, but would see 25 pounds on a scale on the moon and weigh nothing in space. On the other hand, my mass is constant everywhere, unless I change the amount of stuff I am made of (build muscle, get fatter, etc.).

The common weight units in the English System are **Ounces** (oz.) for very light objects (bag of pretzels), **Pounds** (lbs.) for heavier objects (people, bag of potatoes), and **Tons** for extremely heavy objects (elephants, trucks). One pound is 16 ounces, and 1 ton is 2,000 pounds.

Example: Some fire trucks weigh about 20 tons. How many ounces is that?

Solution: Setup unit fractions as

$$20 \; tons \left(\frac{2000 \; lbs}{1 \; ton}\right)\left(\frac{16 \; oz}{1 \; lb}\right)$$

$$= 20 \; \cancel{tons} \left(\frac{2000 \; \cancel{lbs}}{1 \; \cancel{ton}}\right)\left(\frac{16 \; oz}{1 \; \cancel{lb}}\right)$$

$$= 20(2,000)(16) \; oz$$

$$= 640,000 \; oz$$

Try this on your own: Convert 5 gallons into tablespoons.

In the metric system, weights are not typically used, except in science. More often the mass of an object in relation to its weight on Earth is used. The base unit of mass is the **Gram**, abbreviated as "g". One gram is about the amount of metal in a paper clip. It is very small and light.

The same prefixes are used, milli, kilo, etc., with abbreviations such as mg for milligrams and kg for kilograms. On the Earth's surface, an object that weighs 2.2 pounds (rounded), has a mass of one kilogram. This is the conversion factor we will use.

Scientists set the mass units so that one gram is the mass of pure water that has a volume of one milliliter (same as one cubic centimeter). Due to the simple nature of the metric system, one Liter of pure water has a mass of one kilogram.

<u>Example</u>: A common dosage of the medication Thyroxine is 88 milligrams. Convert that to grams.

<u>Solution</u>: Setup unit fractions as

$$88\ mg \left(\frac{1\ g}{1,000\ mg}\right) = 88\ \cancel{mg} \left(\frac{1\ g}{1,000\ \cancel{mg}}\right) = \frac{88}{1,000}\ g = 0.088\ g$$

<u>Example</u>: If a man weighs 180 pounds on Earth, how much mass does he have in kilograms?

<u>Solution</u>: Setup unit fractions as

$$180\ lbs \left(\frac{1\ kg}{2.2\ lb}\right) = 180\ \cancel{lbs} \left(\frac{1\ kg}{2.2\ \cancel{lb}}\right) = \frac{180}{2.2}\ kg = 81.8\ kg$$

In case you are wondering how weight is measured in the metric system, it uses a unit called Newtons (named after the famous scientist Isaac Newton who discovered gravity). Newtons are actually a measure of force, and gravity is a force that attracts objects.

Metric mass has a special unit called a metric **Tonne**, abbreviated as "T", which is 1,000 kilograms. It is slightly larger than an English ton (of course only on Earth). It is equivalent to one megagram (1,000,000 grams), but the word tonne is more commonly used.

Example: How many metric tonnes of mass does a fire trucks weighing 20 tons have?

Solution: Setup unit fractions as

$$20 \, tons \left(\frac{2000 \, lbs}{1 \, ton}\right) \left(\frac{1 \, kg}{2.2 \, lbs}\right) \left(\frac{1 \, T}{1,000 \, kg}\right)$$

$$= 20 \, \cancel{tons} \left(\frac{2000 \, \cancel{lbs}}{1 \, \cancel{ton}}\right) \left(\frac{1 \, \cancel{kg}}{2.2 \, \cancel{lbs}}\right) \left(\frac{1 \, T}{1,000 \, \cancel{kg}}\right)$$

$$= \frac{20(2,000)}{2.2(1,000)} T$$

$$= 18.2 \, T$$

The last topic in this textbook is the relationship between temperature scales. Temperature is a numerical measure of heat intensity. There are different number scales for temperature, most of which are relative scales, meaning that they don't measure the amount of heat, just the relative difference in heat intensity. Hotter objects have higher temperatures and colder objects have lower temperatures.

In the English system, temperature is measured on the **Fahrenheit** scale. This scale is set so that the freezing point of water is 32 degrees and the boiling point of water is 212 degrees. The abbreviation for degrees Fahrenheit is $°F$. Degrees on the Fahrenheit scale can range from -459 (in deep space) up to millions (inside a star).

In the metric system, temperature is measured on the **Celsius** scale. This scale is set so that the freezing point of water is 0 degrees and the boiling point of water is 100 degrees. The abbreviation for degrees Celsius is $°C$. The Celsius scale is sometimes referred to as the

Centigrade scale. Degrees on the Celsius scale can range from -273 (in deep space) up to millions (inside a star).

On a technical note, the freezing and boiling points of water used as reference for temperature scales are measured at sea-level on Earth. They can differ in other places. The minimum temperature is known as <u>Absolute Zero</u>, it is where the molecules essentially stop moving and there is no heat being generated.

There is a simple formula to do conversions between Fahrenheit and Celsius. It is

$$F = \frac{9}{5}C + 32$$

<u>Example</u>: The highest temperature ever recorded at the North Pole is $5°C$. Convert this to Fahrenheit.

<u>Solution</u>: $F = \frac{9}{5}(5°) + 32 = 9 + 32 = 41°F$

<u>Example</u>: Room temperature is defined as $72°F$ (comfortable to the average person). Convert this to Celsius.

<u>Solution</u>: We can setup the equation and then use algebra to rearrange and solve for C.

$$72 = \frac{9}{5}C + 32$$
$$72 - 32 = \frac{9}{5}C$$
$$40\left(\frac{5}{9}\right) = \left(\frac{\cancel{5}}{\cancel{9}}\right)\frac{\cancel{9}}{\cancel{5}}C$$
$$\frac{200}{9} = C$$
$$= 22.2°$$

<u>Try this on your own</u>: Convert $20°F$ to Celsius. Round to whole degree.

Exercises: Converting Units

Solutions appear at the end of this textbook.

1. Convert 237,600 feet to miles.

2. What are the metric prefixes that mean one tenth and one Hundred?

3. Convert 7 kilometers to centimeters.

4. Convert 4 km to yards (round to nearest 0.1).

5. How many square inches in a square yard?

6. Convert 3 square miles to square kilometers.

7. If a property is 1.5 acres, how many square feet is it? How many metric ares?

8. Convert 3 in^3 to mm^3.

9. How many tablespoons in a pint?

10. Convert 12,500 mm^3 to Liters.

11. Convert one metric tonne into ounces.

12. Convert $-10°F$ into Celsius.

Chapter 4

Solutions

Answers to Try This On Your Own Problems

<u>Section 1.1</u>: For the following scenario, describe the population of interest, describe the sample, state the parameter of interest, and the statistic that was calculated. A farm wants to track the weight gain of their chickens after they switched to a new feed. The farm has over 10,000 chickens. They isolated 200 chickens and weighed them before the switch, then every week for the next 10 weeks. At the end of 10 weeks, the 200 isolated chickens gained an average of 1.2 pounds. **ANSWERS: Population is all 10,000 chickens, sample is the 200 isolated chickens, parameter is weight gain, the statistic calculated is 1.2 average weigh gain**

What sampling methods would each description below be classified as?

1. A teacher selected a sample of students by selecting one row and picking all students in that row. **ANSWER: Cluster sampling**

2. A researcher was conducting a survey where they selected a sample by going to every tenth neighborhood and surveying every tenth home from those neighborhoods. **ANSWER: Systematic sampling**

Section 1.2: The grades on a science final exam were 75, 83, 96, 82, 90, 78, 60, 76, 82, 71, 92, 86, 83, 88. Create a table with frequencies and relative frequencies using the intervals 60-69, 70-79, 80-89, 90-99. Then sketch a frequency histogram and a relative frequency pie chart. **ANSWERS:**

Grades	Frequency	Relative Frequency
60-69	1	7%
70-79	4	29%
80-89	6	43%
90-99	3	21%
Total	14	100%

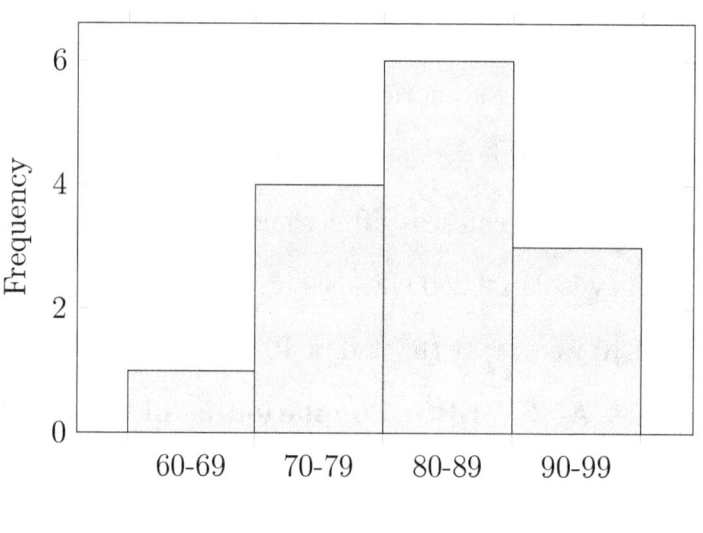

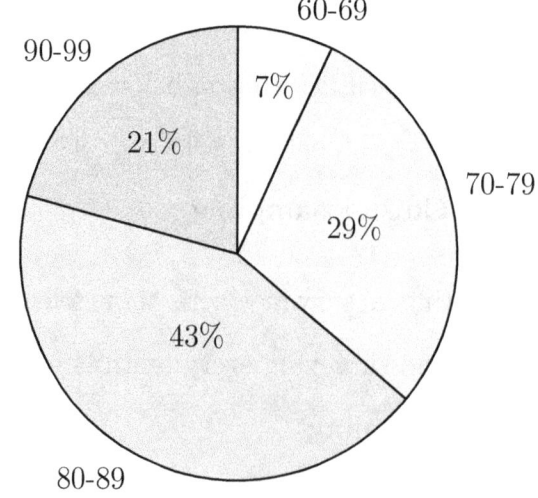

What are the characteristics of the following graph? Examine the spread, symmetry, and outliers.

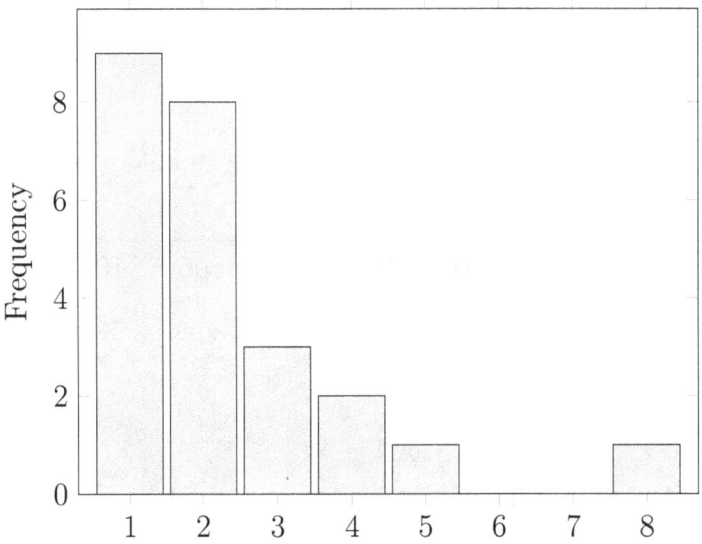

ANSWERS: The peak of the graph is at data values 1 and 2. The outlier is the one value at 8. It is not spread out very much, since most of the data is concentrated near left peak. The graph is right skewed.

Section 1.3: The grades from a sample of a science final exam were 75, 83, 96, 82, 90, 78, 60, 76, 82, 71, 92, 86, 83, 88. Calculate the mean, median, mode, range and standard deviation. **ANSWERS: mean = 81.6, med = 82.5, mode = 82 and 83, range = 36, stddev = 9.3**

Section 1.4: Calculate the z-score of a woman who is 5 feet tall if the mean height is 65 inches and standard deviation is 3 inches. Is she unusually short or not? Round Z to two decimal places. **ANSWER:** $z = \frac{60-65}{3} = -1.67$**, she is a bit short but within the usual values of -2 to 2.**

The grades on a science final exam were 75, 83, 96, 82, 90, 78, 60, 76, 82, 71, 92, 86, 83, 88. Calculate the 5-number summary, IQR, and sketch a regular boxplot. **ANSWERS: min = 60, Q1 = 76, med = 82.5, Q3 = 88, max = 96, IQR = 12**

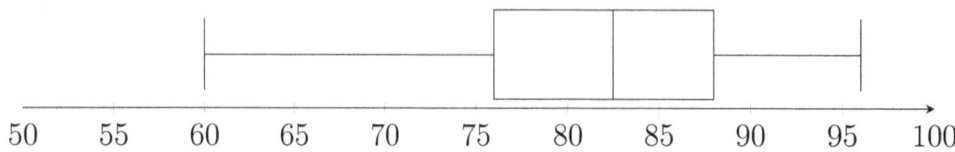

Section 2.1: if $A = \{1, 2, 3, 4\}$, $B = \{3, 2, 4, 1\}$ and $C = \{1, 4, 3\}$, which sets are subsets of the others and are they proper? **ANSWERS: A is a subset of B but not proper, B is a subset of A but not proper, C is a subset of A and a subset of B, it is also a proper subset of both A and B.**

Let the Universe $U = \{1, 2, 3, 4, 5, 6, 7, 8, 9\}$, $A = \{1, 2, 7\}$, and $B = \{2, 4, 6\}$. List the sets $\overline{B}$ and $A \cup B$, then sketch their Venn diagrams. **ANSWERS:** $\overline{B} = \{1, 3, 5, 7, 8, 9\}$ **and** $A \cup B = \{1, 2, 4, 6, 7\}$

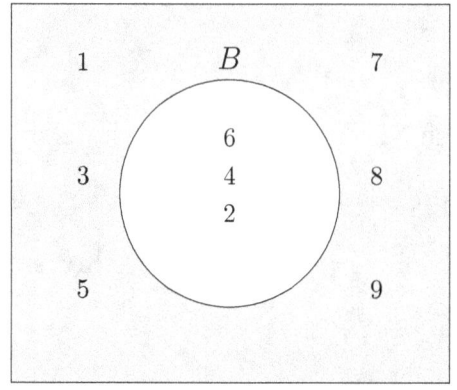

$\overline{B}$ or complement of B

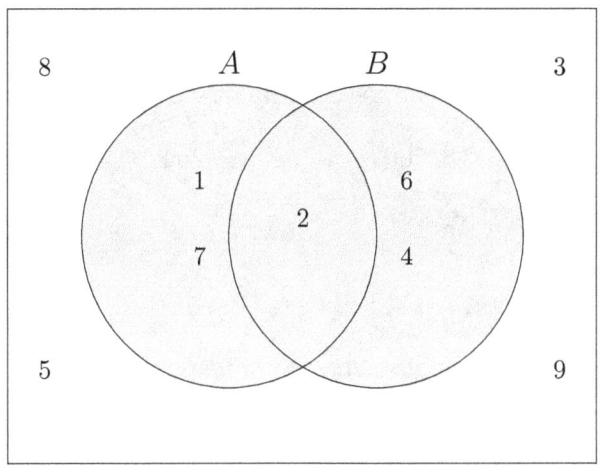

$A \bigcup B$ or union of A or B

Section 2.2: In each situation below, calculate the probability, deciding whether to use empirical or theoretical probability.

1. 200 people are at a banquet and 8 people are at your table including you. What is the probability that someone at your table is chosen at random to win a prize out of the entire banquet? **ANSWER: Theoretical** $P(win) = \frac{8}{200} = 0.04 = 4\%$

2. Danny has played 20 tennis matches this season and has won 17 of them. What is the probability that he wins his next match? **ANSWER: Empirical** $P(win) = \frac{17}{20} = 0.85 = 85\%$

If a team has a 65% chance of winning, find the odds for and against winning. **ANSWERS: P(not win)=35%. Odds for win** $= \frac{65}{35} = \frac{13}{7} =$ **13 to 7. Odds against win** $= \frac{35}{65} =$ **7 to 13.**

Section 2.3: For a lottery in which you pick five numbers from 1 to 50, how many different sets can you pick if they can be in any order, and if they must be in a specific order? **ANSWERS:** $_{50}C_5 = 2,118,760$ **and** $_{50}P_5 = 254,251,200$

Section 2.4: A football team has 42 players. There are 18 players who play offense, 20 players who play defense, and 10 players who play on special teams. Six of the offensive players play both offense and special teams. Find the probability that a player is on the offense or special teams. **ANSWER: P(O or S) = P(O) + P(S) - P(O and S)** = $\frac{22}{42}$ = **0.52 = 52%**

A particular game has the prize distribution shown below. Find the expected value of a prize. **ANSWER: $19**

Prize Amount	$0	$25	$100	$500
Probability	0.7	0.2	0.09	0.01

Section 3.1: Compute $P(-0.5 < Z < 1.35)$ **ANSWER: 0.603 or 60.3%**

The birth weights of babies in Brazil are normally distributed, with a mean of $\mu = 3,110$ grams and a standard deviation of $\sigma = 463$ grams. Find the probability of a baby being born with a weight more than 3,000 grams. **ANSWER: 0.595 or 59.5%**

Section 3.2: The table below shows data from ten people of their average monthly spending on fast food, as well as their average number of days of exercise each month. Create a scatterplot and find the correlation coefficient r. Then find the regression equation and use it to forecast the number of days of exercise output for the fast food value $70. Is that prediction interpolation or extrapolation? Do you notice a pattern between fast food spending and exercise? What could explain the pattern?

fast food $	20	40	58	50	140	30	90	45	100	120
exercise days	20	15	13	11	3	26	7	18	12	1

ANSWER: The scatterplot is below. Overall there seems to be a mildly strong linear pattern in a negative direction. This is confirmed by $r = -0.879$. The regression equation is $y = -0.166x + 24.1$. The prediction is $y = -0.166(70) + 24.1 = 12.5$ days of exercise in a month with $70 fast food spending. This is interpolation, since 70 is in the range of the data (20 to 140). It seems that as fast food spending increases, the number of days of exercise goes down. This could be because people who eat a lot of fast food are not as health conscious as those who eat better, and they would then not exercise as much on average.

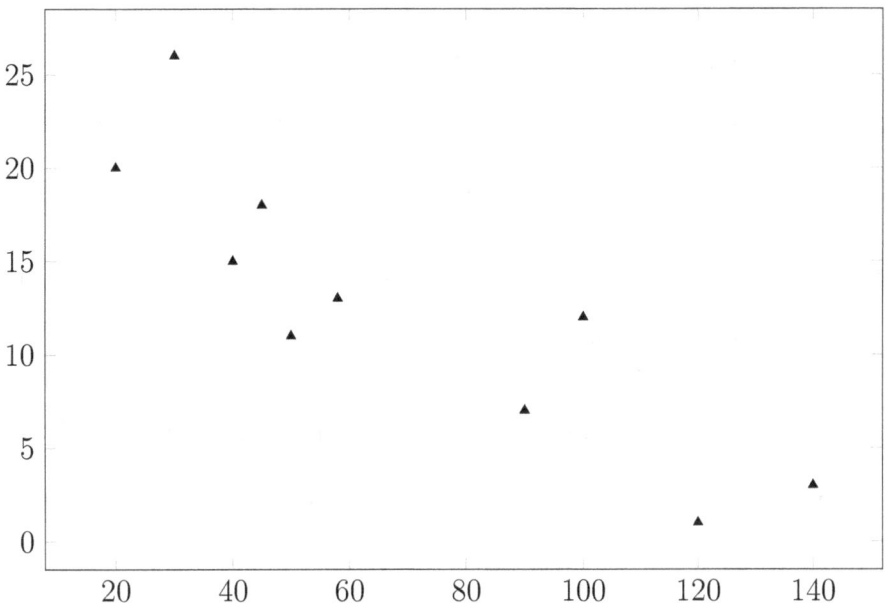

Section 3.3: Convert 7 yards to inches and 310,000 cm to hectometers.

ANSWERS: 252 in. and 31 hm

Convert 8 yards to centimeters using the conversion $1 ft = 30.5\ cm$ ANSWER: $732\ cm$

Convert 5 gallons into tablespoons. ANSWER: 1,280 Tbsp.

Convert $20°F$ to Celsius. Round to whole degree. ANSWER: $-7°C$

Solutions to Exercises: Sec 1.1 Collecting Data

1. The population is all homeschool science textbooks in the United States. The sample is the 15 science books obtained. The (unknown) parameter is the population average price of all the books, which they hope to determine. The statistic measured is the average price of the 15 books = $52.

2. The variables are: name, height, weight, eye color, hair color, and page-hits. The corresponding values are: Sean Higgins, 5ft.10in., 185 lbs., Green, Red, and 142. Name, eye color, and hair color, are qualitative (categories). Height, weight, and number of page-hits are quantitative (measures or counts).

3. Step 1: $m = \frac{6700}{7} = 957.14$, round DOWN to 957. Step2: get random number between 1 and 957 (assumed random value of $k = 121$). Step 3: start at k, keep adding m, to list the place values of the sample selections. Sample is the set of people in places 121, 1078(121+957), 2035, 2992, 3949, 4906, 5863.

4. This is an observational study, since she just observed the crabs and their time. She did not impose treatments or try to control anything.

5. This is stratified sampling, since the items are grouped (strata) and a few from each group are selected. This is not cluster. In cluster, the items are grouped, but entire groups are used.

6. A census is a gathering of information from the entire population of people or things that is being studied. The US government only does a census every ten years, because it takes so much time, money, and staff to complete. Technically, the US census is not a true complete census. It is impossible to keep track of everyone in the country at one exact moment in time. There are people being born and dying every day, criminals or illegal aliens hiding who don't want to be found, and some people who ignore requests or lie about their information.

7. No method is 100% bias free. Simple random is not biased in how it selects, but the sample you get could be biased. Convenience is biased, since it leaves out most of the population. Systematic is biased, because it does not allow most of the combinations to be picked. Cluster can be biased, if the people/items that are together have the same characteristics. Stratified tends to have the least amount of bias. It selects some of each type and the sample is representative of the population.

8. Experimenters use placebos to prevent psychological effects and bias from the experimental units. People who know which treatment they get, can change their stress level and affect the results.

Solutions to Exercises: Sec 1.2 Summarizing Data

1. Since we are trying to summarize, a small number of groups makes it easy to see the big picture. If a data set has 1000 values, using 100 groups would be so large and cumbersome, it would not be a summary and difficult to see anything.

2. When computing relative frequencies, the total should equal 100% or very close. 100% means all the data has been accounted for.

3. A bar graph can be in any order, but a pareto chart has the bars shown in size order. A pareto chart cannot be done from quantitative data, since numbers must go in numerical order of the classes (intervals) and the graph is a histogram.

4. It rained ten days in this month. The distribution of which days it rained would be as follows:

Day	Frequency	Relative Frequency
Tuesday	3	$\frac{3}{10} = 0.3 = 30\%$
Wednesday	1	$\frac{1}{10} = 0.1 = 10\%$
Friday	2	$\frac{2}{10} = 0.2 = 20\%$
Saturday	1	$\frac{1}{10} = 0.1 = 10\%$
Sunday	3	$\frac{3}{10} = 0.3 = 30\%$
Total	10	100%

5. The pie chart and bar graph for which days it rained are:

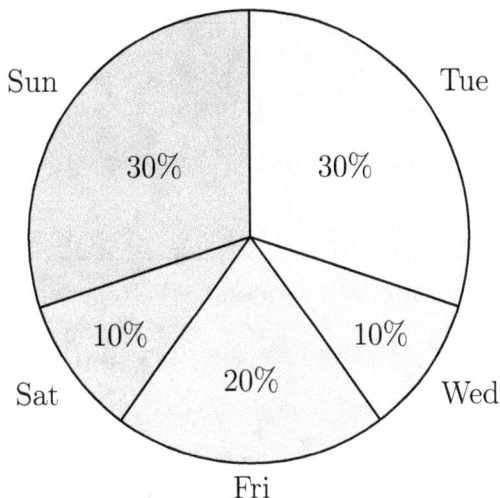

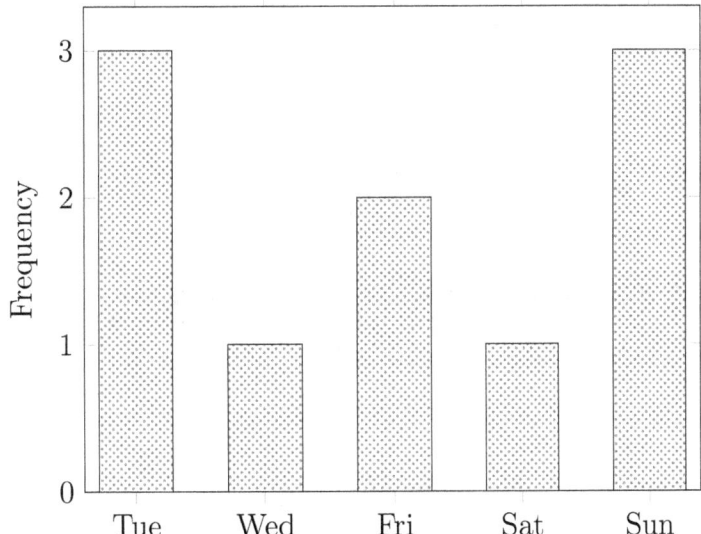

6. For the classes, we can use the common grading scale F (<60), D (60-69), C (70-79), B (80-89), and A (90+). If your school has a different scale, that would be fine also. The distribution is shown below.

Grade	Frequency	Relative Frequency
F (<60)	2	$\frac{2}{19} = 0.10526 = 10.5\%$
D (60-69)	1	$\frac{1}{19} = 0.05263 = 5.3\%$
C (70-79)	4	$\frac{4}{19} = 0.21053 = 21.1\%$
B (80-89)	7	$\frac{7}{19} = 0.36842 = 36.8\%$
A (90+)	5	$\frac{5}{19} = 0.26316 = 26.3\%$
Total	19	100.0%

7. The pie chart and histogram for the history grades are:

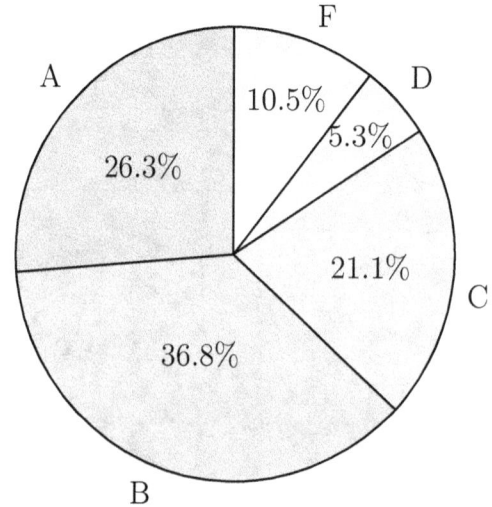

142

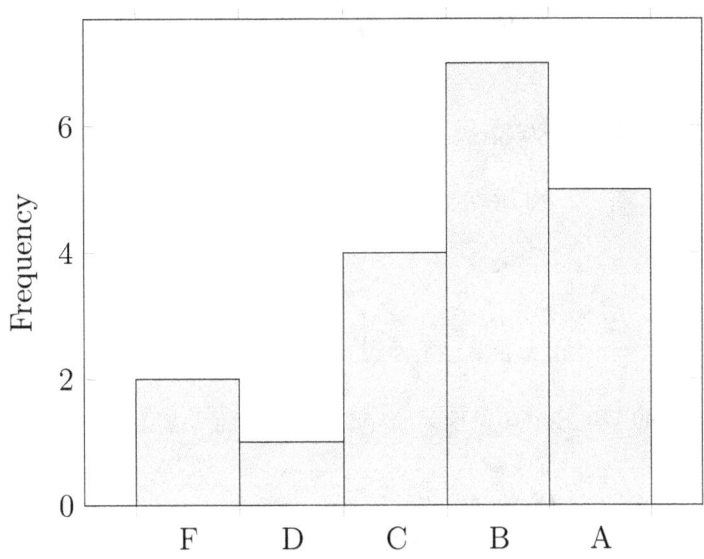

8. The center is at 6. The graph is not spread out that much, it is concentrated from 5-8, with a single outlier at 1. The graph is left-skewed semi-bell shape.

9. The graph has a truncated vertical axis, making the first bar appear much smaller and the third bar much larger than the others. The first bar height around 50, the second bar around 60, not really that much of a difference. The second bar appears to be twice the size of the first, this is misleading. Also there is no title and the categories are vague. This is a very bad graph.

Solutions to Exercises: Sec 1.3 Measuring Data

1. Mean $= \frac{\sum x}{N} = \frac{1527}{19} = 80.3684$, rounded is 80.4. After data is put in order, the median is the 10th value, 82. There are three values which are repeated twice, so the three modes are: 75, 82, and 85.

2. The min is \$44, the max is \$116, so the range is $116 - 44 = \$72$. We can setup a table to help organize the calculations. The mean is 82.

Price (x)	$x - \bar{x}$	$(x - \bar{x})^2$
\$44	$44 - 82 = -38$	$(-38)^2 = 1444$
\$74	$74 - 82 = -8$	$(-8)^2 = 64$
\$94	$94 - 82 = 12$	$12^2 = 144$
\$116	$116 - 82 = 34$	$34^2 = 1156$
Sum		2808

   The variance $s^2 = \frac{\sum (x - \bar{x})^2}{n - 1} = \frac{2808}{4 - 1} = 936$.
   The standard deviation $s = \sqrt{936} = 30.59412$. Using round-off rule, $s = \$30.6$.

3. Notice some are more than \$30.6 away from the average of \$82, and others are less. The ticket prices are spread out by \$30.60 on average away from the mean \$82. This means that the ticket prices vary quite a bit.

4. The median and mode are usually unaffected by extreme values, since the median is in the middle (not at extremes) and the extreme values usually don't occur often. The mean is found from the sum of all values, one extreme value can affect the mean drastically.

5. The two formulas are almost the same, but the sample formula is divided by $n - 1$ instead of just n. The sample standard deviation is always larger than the population standard deviation, because its denominator is less, making the fraction more.

6. $GPA = \dfrac{3.0(3) + 3.0(4) + 4.0(2) + 2.0(3) + 4.0(3)}{3 + 4 + 2 + 3 + 3} = \dfrac{47}{15} = 3.13$, which is a low B, not bad for a first semester in college.

7. Since the mean is greater than the median, this data set is probably right-skewed. The top half of the weights are spread far out into very large values. This matches with the large standard deviation. There must be some women well more than 45 pounds over average.

8. The midpoints of the classes ($\widehat{x}$) are: $12, 17, 22, 27, 32, 37$. Then the formula would be
$\dfrac{\sum \widehat{x} f}{\sum f} = \dfrac{12(12)+17(5)+22(7)+27(2)+32(6)+37(3)}{12+5+7+2+6+3} = \dfrac{740}{35} = 21.14286$, rounded to 21.1.

9. For the shot put, $CV = \dfrac{5.5}{38} \times 100\% = 14.5\%$.
   For the gymnastics, $CV = \dfrac{1.4}{8.45} \times 100\% = 16.6\%$.
   The gymnastics scores are a more spread out set of data than the shot put throws.

Solutions to Exercises: Sec 1.4 Measuring of Relative Standing

1. The z-scores are shown in the table below.

Name	IQ score	z-score
Garry Kasporov	190	$z = \frac{190-100}{15} = +6.00$
Albert Einstein	160	$z = \frac{160-100}{15} = +4.00$
Arnold Schwarzenegger	135	$z = \frac{135-100}{15} = +2.33$
Tim Tebow	104	$z = \frac{104-100}{15} = +0.27$
Howard Stern	99	$z = \frac{99-100}{15} = -0.07$
George W Bush	125	$z = \frac{125-100}{15} = +1.67$
Muhammad Ali	78	$z = \frac{78-100}{15} = -1.47$
Barack Obama	130	$z = \frac{130-100}{15} = +2.00$

   Usual values are between -2 and $+2$, so the only ones that are unusual are Kasporov, Einstein, and Schwarzenegger. Note: Kasporov is way off the charts, among the smartest humans ever.

2. Unusually short would be a z-score below -2, so we can setup the formula and solve for $x < 63$ inches (or 5ft 3in).

$$z = \frac{x - \mu}{\sigma}$$
$$-2 = \frac{x - 69}{3}$$
$$3(-2) = \left(\frac{x - 69}{3}\right)3$$
$$-6 = x - 69$$
$$69 - 6 = x$$

Unusually tall would be a z-score above +2, so we can setup the formula and solve for $x > 75$ inches (or 6ft 3in).

$$z = \frac{x - \mu}{\sigma}$$
$$2 = \frac{x - 69}{3}$$
$$3(2) = \left(\frac{x - 69}{\cancel{3}}\right)\cancel{3}$$
$$6 = x - 69$$
$$69 + 6 = x$$

3. $L = 38\left(\frac{45}{100}\right) = 17.1$, so the percentile is the data value in the 17th position. We don't have the data set here, so we cannot state the actual value, it would just be whatever value is in that location of 38 values put in order.

4. It means that their score on the test was greater than (or equal to) 85% of all the scores. They beat 85% of the test takers.

5. First put the data values in size order. The tenth value is in the middle (median). Q_1 is the median of the first 9 values and Q_3 is the median of the last 9 values. The five-number summary is Min $= 48, Q_1 = 75$, Med $= 82, Q_3 = 90$, Max $= 100$.

6. The boxplot looks like:

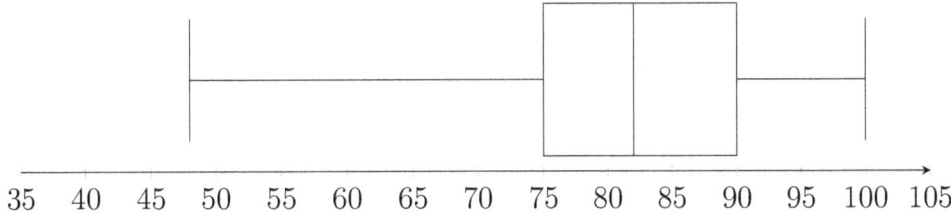

7. $IQR = 90 - 75 = 15$, so $LF = 75 - 1.5(15) = 52.5$ and $UF = 90 + 1.5(15) = 112.5$. There are two outliers (48 and 50) on the low end (below 52.5), but no outliers on the high end (above 112.5). Then modified boxplot looks like:

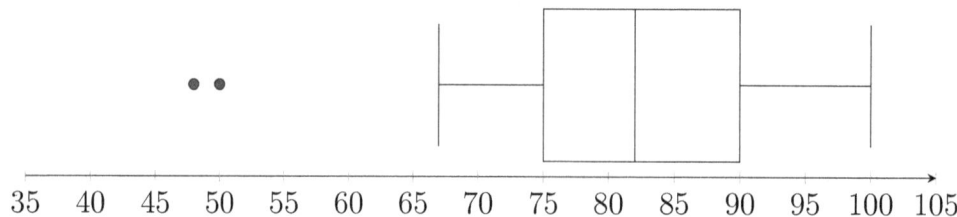

8. Group C is the most spread out, the box is wider and the whiskers extend out farther that the other boxplots. Group C is symmetric, Group B is left-skewed, and Group A is right-skewed.

Solutions to Exercises: Sec 2.1 Sets and Venn Diagrams

1. For a phone number of $(212)555-1239$, $N = \{1, 2, 3, 5, 9\}$ and $\overline{N} = \{0, 4, 6, 7, 8\}$.

2. $E = A$, since they both have the same elements. No other sets are equal. Technically, each set is a subset of itself. D is a proper subset of B. C is a proper subset of A, E, and B. G is a proper subset of O and B.

3. $V \bigcap B = \{a, e\}$, $B \bigcap R = \{\}$ (or the empty set $\emptyset$), and $V \bigcup R = \{a, e, i, o, u, j, q, x, z\}$

4. The Venn diagram for $V \bigcap B$ is:

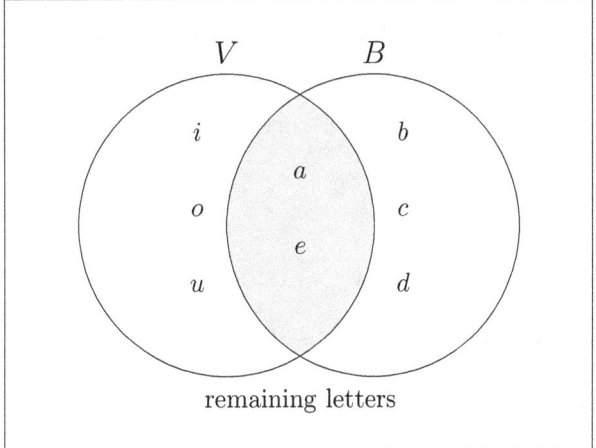

The Venn diagram for $B \bigcap R$ is below. Notice no shading, since the intersection is the empty set.

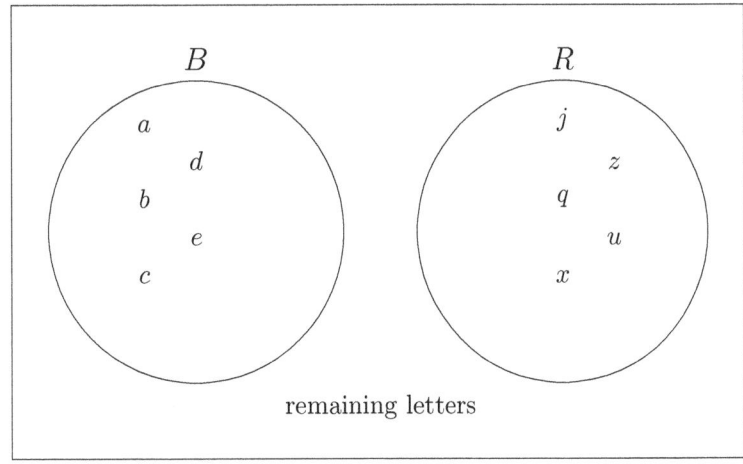

149

The Venn diagram for $V \bigcup R$ is:

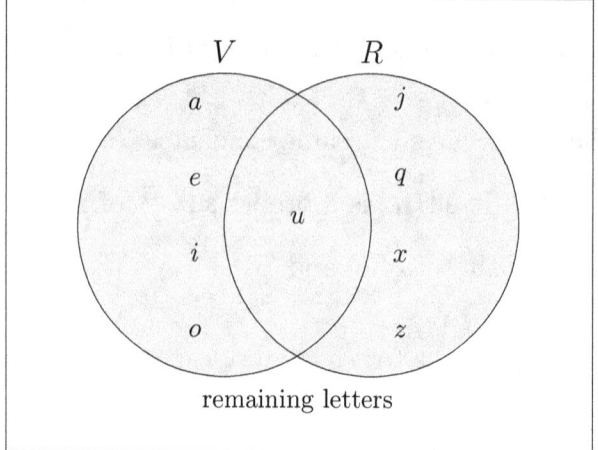

5. Your set could be something like C = {geometry, world lit, biology, art, spanish}

6. The shaded covers the elements that are inside set E and at the same time, outside of set H. So this is $E \bigcap \overline{H}$.

7. $\overline{A} = \{6, 7, 8, 9\}$ which is everything outside of the A circle.

   First we find the intersection $A \bigcap B = \{2, 4\}$, then its complement is all other numbers $\overline{A \bigcap B} = \{1, 3, 5, 6, 7, 8, 9\}$.

   The complement of B is $\overline{B} = \{1, 3, 5, 7, 9\}$, then union with set A gives $A \bigcup \overline{B} = \{1, 2, 3, 4, 5, 7, 9\}$.

Solutions to Exercises: Sec 2.2 Probability Basics

1. List the sample spaces for the following experiments:

   (a) SS = { H, T }, $N = 2$.

   (b) We need to look at what happens for each flip and combine them.

   SS = { HHH, HHT, HTH, HTT, THH, THT, TTH, TTT }, $N = 8$.

   (c) $SS = \{1, 2, 3, 4, 5, 6\}$, $N = 6$.

   (d) The sums range from 2 (rolling a 1+1) up to 12 (rolling 6+6)

   $SS = \{2, 3, 4, 5, 6, 7, 8, 9, 10, 11, 12\}$, $N = 11$.

   (e) Using letter abbreviations for the 7 rainbow colors and seasons,

   SS = { Rw, Ow, Yw, Bw, Gw, Iw, Vw, Rsp, ..., Rsm, ..., Rf, Of, Yf, Bf, Gf, If, Vf }, $N = 28$.

2. Probabilities must be between 0 and 1 (or 0% and 100%), so the valid values are 0.35, 0.004, and $\frac{3}{8}$. All of the others are either negative or greater than 1.

3. The event 'even' consists of the outcomes 2, 4, and 6. This is three out of the 6 possible outcomes, so $P(even) = \frac{3}{6} = 0.5 = 50\%$. $P(3) = \frac{1}{6} = 0.167 = 17\%$. P(>2) = P(3 or 4 or 5 or 6) = $\frac{4}{6} = 0.667 = 67\%$.

4. Probability of making a shot, based on his data, is $\frac{5}{12} = 0.417 = 42\%$. Subjectively, you might believe this to be low percentage and he might not make the team.

5. The probability of precipitation equals 0.45 (the sum of all three given). By the complement rule, the probability of no precipitation = 1 - probability of precipitation $= 1 - 0.45 = 0.55 = 55\%$.

6. The probability of no rain today is $100\% - 20\% = 80\%$. Odds in favor $= \dfrac{P(rain)}{P(no\ rain)} = \dfrac{20}{80} = \dfrac{1}{4}$ or 1 to 4 odds in favor. Odds against rain $= \dfrac{P(no\ rain)}{P(rain)} = \dfrac{80}{20} = \dfrac{4}{1}$ or 4 to 1 odds against.

7. The theoretical probabilities of rolling each number are all equal to $\frac{1}{6}$ or about 17%. Your empirical probabilities are computed by dividing the count of how often a number was rolled, by the total of 15. Your values most likely vary, some smaller than 17% and some larger. If they are close to the theoretical, you got lucky. If they are not, that is because your rolls are random and with only 15 rolls, the law of large numbers does not work very well.

Solutions to Exercises: Sec 2.3 Counting Rules

1. Using the Fundamental Counting Principle, the number of combo meals is

   $5 \times 4 \times 3 = 60$.

2. $_8C_3 = \frac{8!}{3!(8-3)!} = \frac{8 \cdot 7 \cdot 6 \cdot 5 \cdot 4 \cdot 3 \cdot 2 \cdot 1}{3 \cdot 2 \cdot 1 (5 \cdot 4 \cdot 3 \cdot 2 \cdot 1)} = \frac{8 \cdot 7 \cdot 6}{3 \cdot 2 \cdot 1} = 56$

   $_{11}C_9 = \frac{11!}{9!(11-9)!} = \frac{11 \cdot 10 \cdot 9 \cdot 8 \cdot 7 \cdot 6 \cdot 5 \cdot 4 \cdot 3 \cdot 2 \cdot 1}{9 \cdot 8 \cdot 7 \cdot 6 \cdot 5 \cdot 4 \cdot 3 \cdot 2 \cdot 1 (2 \cdot 1)} = \frac{11 \cdot 10}{2 \cdot 1} = 55$

   $_7P_4 = \frac{7!}{(7-4)!} = \frac{7 \cdot 6 \cdot 5 \cdot 4 \cdot 3 \cdot 2 \cdot 1}{3 \cdot 2 \cdot 1} = 7 \cdot 6 \cdot 5 \cdot 4 = 840$

   $_8P_8 = \frac{8!}{(8-8)!} = \frac{8!}{0!} = \frac{8!}{1} = 8 \cdot 7 \cdot 6 \cdot 5 \cdot 4 \cdot 3 \cdot 2 \cdot 1 = 40,320$

   $_5C_1 = \frac{5!}{1!(5-1)!} = \frac{5!}{4!} = \frac{5 \cdot 4 \cdot 3 \cdot 2 \cdot 1}{4 \cdot 3 \cdot 2 \cdot 1} = 5$

3. If we consider sequences of the 3 movies, that implies a specific order, so we want permutation here. $_{10}P_3 = \frac{10!}{(10-3)!} = \frac{10!}{7!} = \frac{10 \cdot 9 \cdot 8 \cdot 7 \cdot 6 \cdot 5 \cdot 4 \cdot 3 \cdot 2 \cdot 1}{7 \cdot 6 \cdot 5 \cdot 4 \cdot 3 \cdot 2 \cdot 1} = 10 \cdot 9 \cdot 8 = 720$

4. Since we are just looking at the numbers selected, and not in any order, this is a combination. $_{39}C_5 = \frac{39!}{5!(39-5)!} = \frac{39 \cdot 38 \cdot 37 \cdot 36 \cdot 35}{5 \cdot 4 \cdot 3 \cdot 2 \cdot 1} = \frac{69090840}{120} = 575,757$ different sets of 5 numbers to play.

5. The probability of winning the Fantasy 5 lottery jackpot (matching all 5 numbers) is $\frac{1}{575757} = 0.00000174$, pretty small chance. There is one jackpot winning set of numbers and 575,756 losing sets, so the odds against winning are 575,756 to 1, very much stacked against you!

6. It is impossible to compute $_5C_9$, since this means to select 9 items from a group of 5. This makes no sense. Also the formula would have a negative value, and factorial is only for positive values.

7. We need to get a count of how many combinations with 3 matching numbers, and how many combinations there are in total, then divide them. If we select 3 winning numbers, that means we also have selected 2 losing numbers. By the Fundamental Counting Principle, we multiply the number of ways to pick 3 out of 5 winning numbers, by the

number of ways to pick 2 out of the 24 losing numbers. Each of these is a combination, so our numerator is $_5C_3 \times {}_{24}C_2$. The total number of sets of 5 is $_{29}C_5$. Now the probability is $\dfrac{_5C_3 \times {}_{24}C_2}{_{29}C_5} = \dfrac{10(276)}{118755} = 0.023 = 2.3\%$

Solutions to Exercises: Sec 2.4 More Probability

1. Mutually exclusive means that the events cannot happen at the same time. One example is rolling a 3 on a die and a 5 on a die. Another example would be the experiment picking a name for a raffle winner, with events picking female and picking male.

2. We add up all of the physics or engineering majors, but make sure not to double count the 14 double majors. There are $32 + (112 - 14) = 130$ in that group. For total students we add all three majors without double counting, $32 + (49 - 8) + (112 - 14) = 171$. Now P(Phys or Eng)$= \frac{130}{171} = 0.760 = 76\%$.

3. Using addition rule, $P(A \text{ or } B) = P(A) + P(B) - P(A \text{ and } B) = 0.5 + 0.7 - 0.3 = 0.9$.

4. Setting up the addition rule, $P(A \text{ or } B) = P(A) + P(B) - P(A \text{ and } B)$, we get $0.85 = 0.65 + P(B) - 0.25$. By simplifying and solving, we get $P(B) = 0.45$

5. These are dependent events, the first card picked affects what cards are left and so affects the second pick. We use the multiplication rule $P(\text{red6 and red6}) = P(\text{1st red6}) \cdot P(\text{2nd red6 | 1st red6}) = \frac{2}{52} \cdot \frac{1}{51} = \frac{1}{1326} = 0.00075 = 0.075\%$, which is a very small chance.

6. By reasoning, we know the ten of hearts is one out of 26 red cards. By formula, $P(\text{10hearts|red}) = \dfrac{P(\text{10hearts and red})}{P(\text{red})} = \dfrac{\frac{1}{52}}{\frac{26}{52}} = \dfrac{1}{26} = 0.038 = 3.8\%$

7. One example is rolling two dice and getting a 3 on one die and a 5 on the other die. Another example is randomly picking a person and determining that they work at Home Depot and like vanilla ice cream. Where you work has no affect on what ice cream you like.

8. Two events can never be both mutually exclusive and independent. If they are mutually exclusive, the occurrence of one automatically affects the other (it prohibits the other). Once you know a coin lands heads up, then tails has no chance (until the next flip).

9. This is not a valid probability distribution. The probabilities themselves are valid, but the total only sums to 0.97 or 97%.

10. $E(x) = \frac{752,944}{800,000}(0) + \frac{45,000}{800,000}(5) + \frac{2,000}{800,000}(100) + \frac{50}{800,000}(2500) + \frac{5}{800,000}(20000) + \frac{1}{800,000}(100000) = \$0.94$, so the lottery expects to pay out $0.94 on average for each play of the game. In order to make a profit, they need to charge more than that, maybe $1.00 or $2.00.

Solutions to Exercises: Sec 3.1 The Normal Distribution

1. We are looking for $P(Z > -1.04)$. Draw a bell curve, make a line to slice the graph at $z = -1.04$, shade above (to right), then go to the table at end of the book. Look on the first page (negative z-scores) and go down to the row for -1.0 and across to the fifth column 0.04. There we find the area of 0.1492, which is the area to the left, but we want the area to the right. Therefore, $P(Z > -1.04) = 1 - P(Z < -1.04) = 1 - 0.1492 = 0.8508 = 85.08\%$. The graph is shown below. Notice that the probability is very large and the shaded area is very large, so this makes sense.

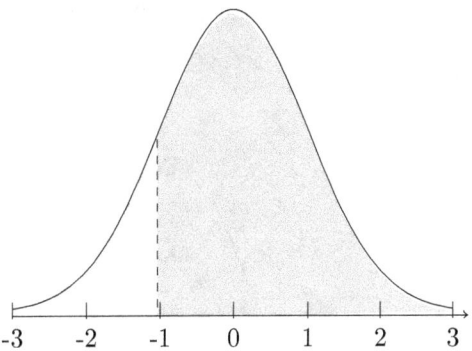

2. Find $P(Z < 2.73)$. Draw a bell curve, make a line to slice the graph at $z = 2.73$, shade below (to left), then go to the table at end of the book. Look on the second page (positive z-scores) and go down to the row for 2.7 and across to the first column 0.03. There we find the area of 0.9968, which is our answer, $P(Z < 2.73) = 0.9968 = 99.68\%$. The graph is shown below.

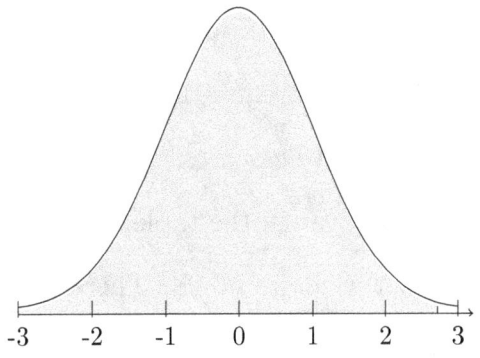

3. We are looking for $P(-1 < Z < 2)$. Draw a bell curve, make lines to slice the graph at $z = -1$ and $z = 2$, shade above that range, then go to the table at end of the book. Look on the first page (negative z-scores) and go down to the row for -1.0 and across to the first column 0.00 (since the z-score is really -1.00). There we find the area of 0.1587, which is the area to the left of $z = -1$. Look on the second page (positive z-scores) and go down to the row for 2.0 and across to the first column under 0.00. There we find the area of 0.9772, which is the area to the left of $z = 2$. Now we want the area between these, so we subtract these areas. Therefore, $P(-1 < Z < 2) = P(Z < 2) - P(Z < -1) = 0.9772 - 0.1587 = 0.8185 = 81.85\%$. The graph is shown below. Notice that the size of the probability matches the size of the shaded area, so this makes sense.

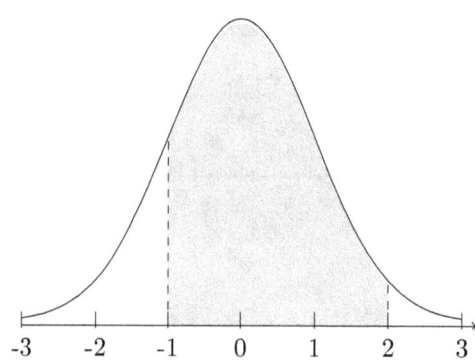

4. We want the z, such that $P(Z > z) = 0.0700$. Draw a bell curve, draw a line to slice the graph into a somewhat small slice over to the right, shade above (to right), then go to the table at end of the book. Since the table shows areas to the left (below), and our given area is to the right (above), we need to use the complement rule to convert, in order to match the table. The area to the left is $1 - 0.0700 = 0.9300$ (or 93%). Look on the second page with positive z-scores. Look in the body of the table for 0.9300. This exact value does not appear in the table, but the closest value is 0.9306. It is in the row for 1.4 and below the column 0.08. Therefore, the z-score is 1.48. The graph is shown on the next page.

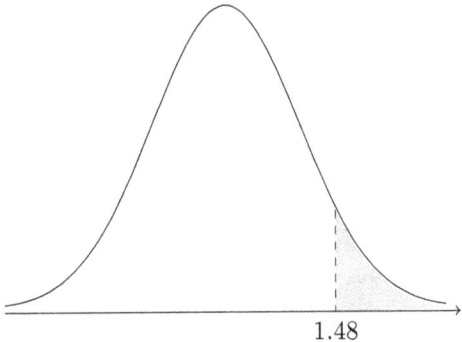
1.48

5. With 10% on each edge, there is 80% in the middle section. We want z_1 and z_2, such that $P(z_1 < Z < z_2) = 0.80$. Draw a bell curve, draw a line to slice the graph into mirror image small slices, one over to the right and one over to left, shade the outer edge slices beyond them, then go to the table at end of the book. The area to the left of the low edge is 0.10. Look on the first page, negative z-scores. Look in the body of the table for 0.1000. This exact value does not appear in the table, but the closest value is 0.1003. It is in the row for -1.2 and below the column 0.08. Therefore, the z-score is $z_1 = -1.28$. Because of the perfect symmetry, the positive cutoff on the other edge is $z_2 = 1.28$. The graph is shown below.

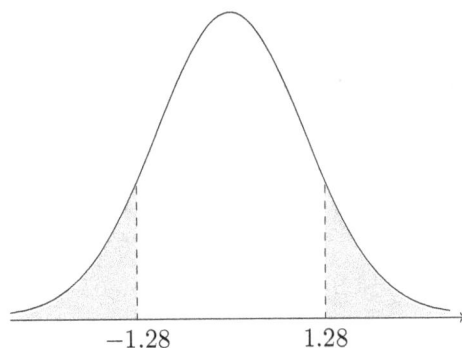
-1.28 1.28

6. This is due to the symmetry of the graph. Going from a certain value below zero and over to the right, is a mirror image of its absolute value and over to the left.

7. Draw a bell curve with standard z-axis from -3 to 3 and below that, an x-axis with heights that correspond to the z marks. The mean height of 64 will go below $z = 0$, one standard deviation higher (66.5 inches) will go below $z = 1$, two deviations higher (69)

goes below $z = 2$, and 71.5 below $z = 3$. Do similar process on left side, subtracting standard deviation to go under the negative z-values. Change 5 feet 6 inches into 66 inches. Then convert 66 into a z-score, $z = \frac{66-64}{2.5} = 0.80$. Make a line to slice the graph at about $z = 0.80$, shade below (to left), then go to the table at end of the book. Look on the second page (positive z-scores) and go down to the row for 0.8 and across to the first column 0.00. There we find the area of 0.7881, which is the area to the left, exactly what we need. Therefore, $P(X < 66) = P(Z < 0.80) = 0.7881 = 78.81\%$. A bit more than $\frac{3}{4}$ of women are shorter than 5 foot 6 inches. The graph is shown below.

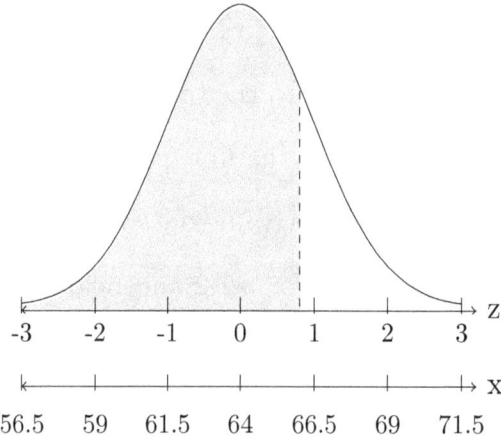

8. The birth weights of hospital born babies in Pakistan are normally distributed, with a mean of $\mu = 2.9$ kg and a standard deviation of $\sigma = 0.5$ kg. Find the probability of a baby being born with a weight greater than 3.5 kg. Draw a bell curve with standard z-axis from -3 to 3 and below that, an x-axis with weights that correspond to the z marks. The mean weight of 2.9 will go below $z = 0$, one standard deviation higher (3.4 kg) will go below $z = 1$, etc. Convert 3.5 into a z-score, $z = \frac{3.5-2.9}{0.5} = 1.20$. Make a line to slice the graph at about $z = 1.20$, shade above (to right), then go to the table at end of the book. Look on the positive z-scores page and go down to the row for 1.2 and across to the first column 0.00. There we find the area of 0.8849, which is the area to the left, but we want the area to the right. Therefore, $P(X > 3.5) = P(Z > 1.2) = 1 - P(Z < 1.2) = 1 - 0.8849 = 0.1151 = 11.51\%$. A

bit more than 11% of babies in the Pakistan are born very large. The graph is shown below.

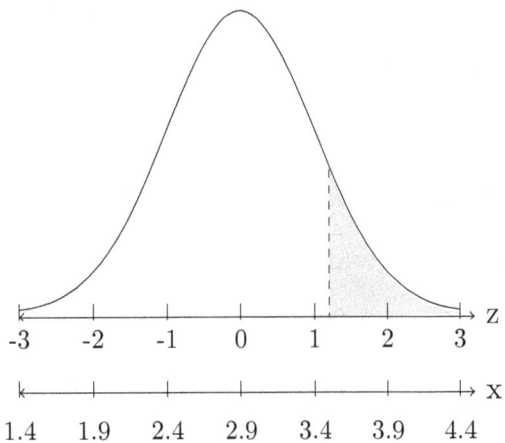

9. Draw a bell curve, draw a line to slice the graph into a small slice way over to the right, shade below (to left), then go to the table at end of the book. Look on the positive z-score page and look in the body of the table for 0.7500. This exact value does not appear in the table, but the closest value is 0.7486. It is in the row for 0.6 and below the column 0.07. Therefore, the z-score is 0.67. Using the formula $z = \frac{x-\mu}{\sigma}$, we can solve for the unknown score x. Formula setup is $0.67 = \frac{x-500}{100}$. After multiplying both sides by 100 and adding 500, we get $x = 567$ as the 90th percentile. This means that 75% of all people score lower than 567 on any section of the SAT. The graph is shown below.

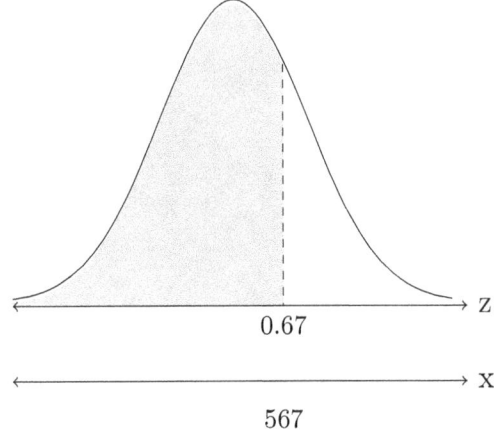

10. Draw a bell curve, draw a line to slice the graph into a small slices way over to the right, shade between, then go to the table at end of the book. Look on the positive z-score page and look in the body of the table for 0.8500 and 0.9500. These exact values do not appear in the table, but the closest values are 0.8508 and 0.9505 or 0.9495. The first one is in the row for 1.0 and below the column 0.04. Therefore, the z-score is 1.04. The other is exactly halfway between, so the average of the corresponding z-scores is $z = 1.645$. Using the z-score formula for each, we get $x = 604$ as the 85th percentile and $x = 665$ as the 95th percentile. This means that applicants who score between 604 and 664 on the math section of the SAT will qualify. The graph is shown below.

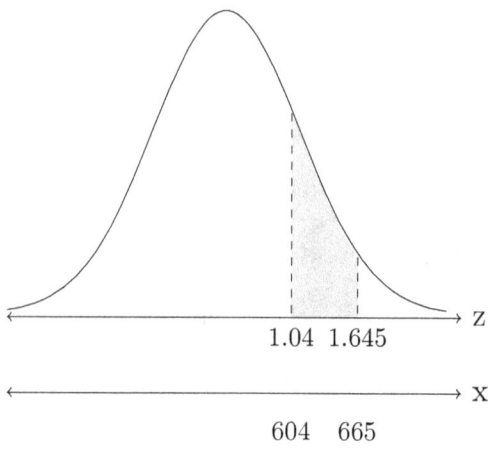

Solutions to Exercises: Sec 3.2 Correlation and Regression

1. Correlation is when two variables tend to be related. Certain values of one variable tend to be paired with certain values of the other. For positive correlation, as one goes up (increases), the other tends to go up. For negative correlation, as one goes up, the other tends to go down (decreases).

2. Match the most likely linear correlation values to the graphs below.
   $r = +0.7 \Rightarrow b$ $r = +0.99 \Rightarrow a$ $r = -0.4 \Rightarrow c$ $r = +0.15 \Rightarrow e$ $r = -0.86 \Rightarrow d$
   $r = 0 \Rightarrow f$

3. We can add extra rows to organize the calculations.

x	2	5	7	10	12	14	$\sum x = 50$
y	2	6	7	9	11	14	$\sum y = 49$
x^2	4	25	49	100	144	196	$\sum x^2 = 518$
y^2	4	36	49	81	121	196	$\sum y^2 = 487$
xy	4	30	49	90	132	196	$\sum xy = 501$

   Now we input the value $n = 6$ and all of the sums into the formula to get:

   $$r = \frac{501 - \frac{50(49)}{6}}{\sqrt{\left[518 - \frac{(50)^2}{6}\right]\left[487 - \frac{(49)^2}{6}\right]}} = \frac{501 - 408.33}{\sqrt{[518 - 416.67][487 - 400.17]}} = \frac{92.67}{93.80} = +0.988$$

4. Go to STAT EDIT menu, enter the SAT data into L_1. Then move over to L_2 and input the GPA data exactly in order. Press STAT, CALC menu, scroll down to item LinReg(ax+b) and press ENTER. Type L_1, L_2 then hit enter again. The results you should see are $r = +0.793$ (rounded). This is greater than the minimum value from the table of 0.754, therefore, the value $r = +0.793$ is large enough to state that there is a strong correlation between SAT math score and college GPA. It is probably that

the students who get higher scores, work harder, study well, and are more interested in learning, so they do well in college.

To get the scatterplot, press 2nd Y= (for stat plot menu), choose Plot1, select first type icon for scatterplot. Set the Xlist to L_1 and Ylist to L_2, then press GRAPH . If you do not see it, go to ZOOM and select ZoomStat, hit enter. The scatterplot will look like this:

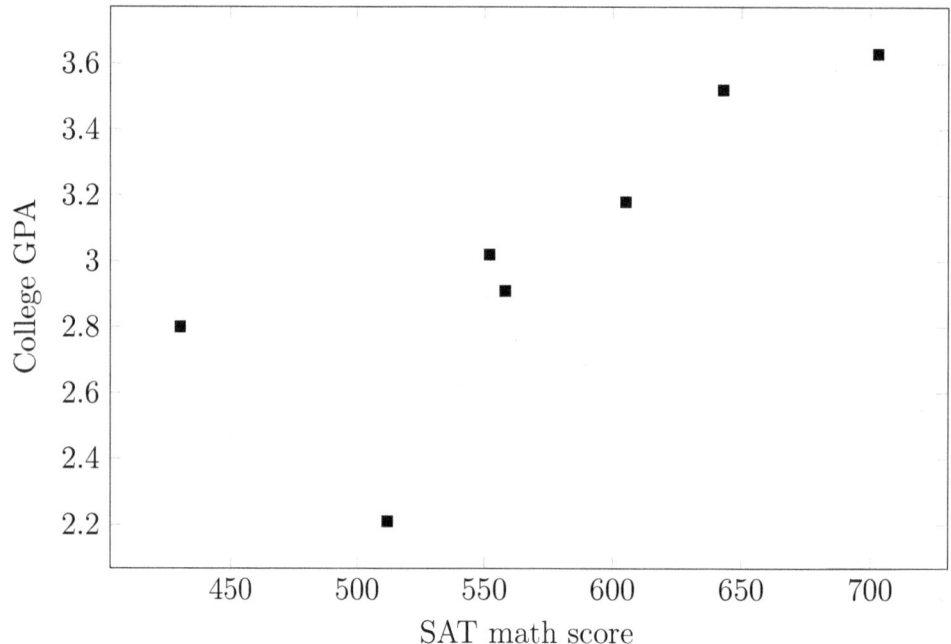

5. Press STAT , CALC menu, scroll down to item LinReg(ax+b) and press ENTER. Type L_1, L_2 then hit enter again. The results you should see are $y = 0.004x + 0.613$ (rounded). These values imply that for every point your SAT math score increases, your college GPA will increase by 0.004, and that with a zero on SAT math, you should still be able to get a 0.613 GPA. In reality, the lowest SAT score is 200, but the equation follows that pattern if we could project down to zero. The regression line is shown along the scatterplot points below.

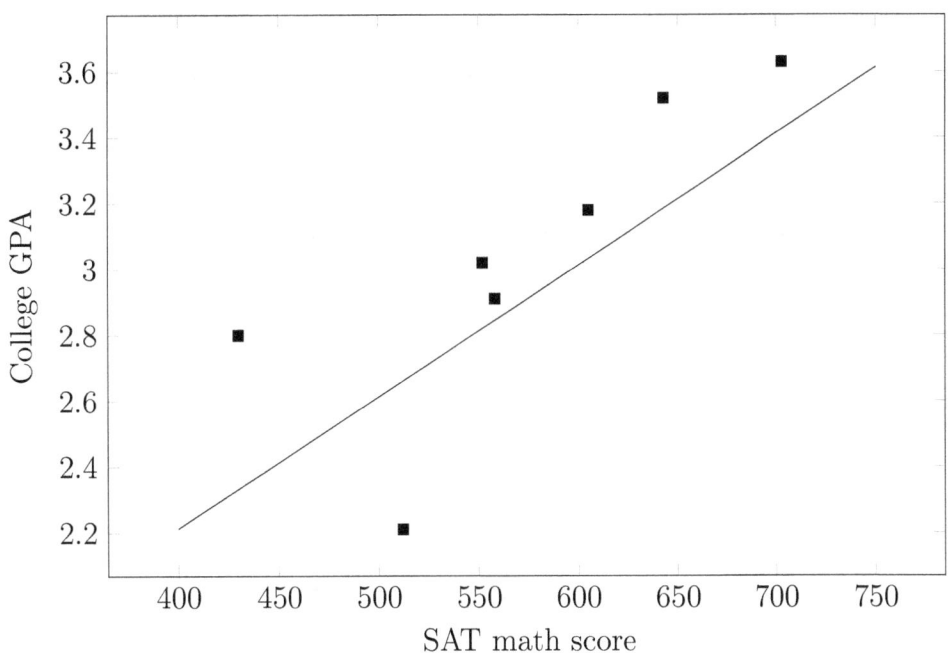

6. The predictions are $y = 0.004(760) + 0.613 = 3.653$ GPA with an SAT math score of 760, and $y = 0.004(500) + 0.613 = 2.613$ GPA with an SAT math score of 500. The first one is extrapolation, since 760 is outside the range of the data (430 to 703). The second is interpolation, since 500 is in the range.

7. Two possible regression lines are shown below. Other similar lines would be good fits as well. There appears to be a somewhat strong negative correlation between the variables. More TV watching tends to match with lower scores. It is reasonable to assume that watching a lot of TV, takes time away from studying, and therefore causes lower scores (on average). There are exceptions, but this is a general fact backed up by much research.

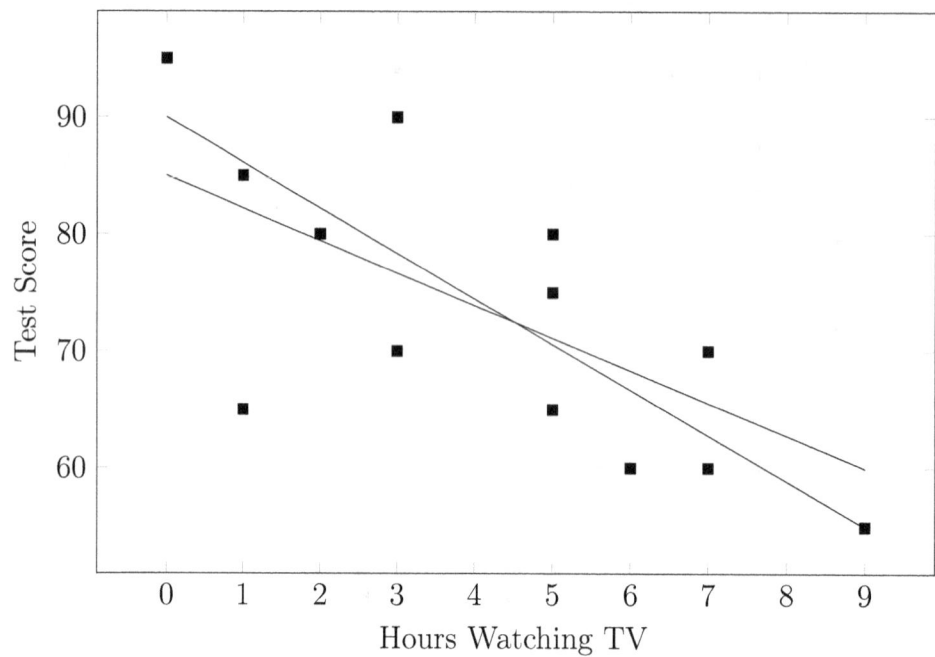

8. A positive correlation coefficient r corresponds with a positive slope. A negative correlation coefficient r corresponds with a negative slope. The values have no relation, only the sign, $\pm$.

Solutions to Exercises: Sec 3.3 Converting Units

1. $237,600 \, ft \left(\dfrac{1 \, mile}{5,280 \, ft} \right) = \dfrac{237,600}{5,280} = 45 \, miles$

2. Deci means one tenth and Hecto means one Hundred.

3. $7 \, km \left(\dfrac{1,000 \, m}{1 \, km} \right) \left(\dfrac{100 \, cm}{1 \, m} \right) = 7(1,000)(100) = 700,000 \, cm$

4. Multiply by unit fractions to get to centimeters, then over to inches, then yards. Yards are much smaller than kilometers, so the answer should be a much larger number.

$$4 \, km \left(\dfrac{1,000 \, m}{1 \, km} \right) \left(\dfrac{100 \, cm}{1 \, m} \right) \left(\dfrac{1 \, in}{2.54 \, cm} \right) \left(\dfrac{1 \, ft}{12 \, in} \right) \left(\dfrac{1 \, yd}{3 \, ft} \right)$$

$$= 4 \, \cancel{km} \left(\dfrac{1,000 \, \cancel{m}}{1 \, \cancel{km}} \right) \left(\dfrac{100 \, \cancel{cm}}{1 \, \cancel{m}} \right) \left(\dfrac{1 \, \cancel{in}}{2.54 \, \cancel{cm}} \right) \left(\dfrac{1 \, \cancel{ft}}{12 \, \cancel{in}} \right) \left(\dfrac{1 \, yd}{3 \, \cancel{ft}} \right)$$

$$= \dfrac{4(1,000)(100)}{2.54(12)(3)} yds$$

$$= 4,374.5 \, yds$$

5. One square yard is $3 \, ft \times 3 \, ft = 36 \, in \times 36 \, in = 1,296 \, in^2$.

6. Multiply by unit fractions to get to feet and inches, then convert over to centimeters, then to kilometers.

$$3 \, miles^2 \left(\dfrac{5,280 \, ft}{1 \, mile} \right)^2 \left(\dfrac{12 \, in}{1 \, ft} \right)^2 \left(\dfrac{2.54 \, cm}{1 \, in} \right)^2 \left(\dfrac{1 \, m}{100 \, cm} \right)^2 \left(\dfrac{1 \, km}{1,000 \, m} \right)^2$$

$$= 3 \, \cancel{miles^2} \left(\dfrac{27,878,400 \, \cancel{ft^2}}{1 \, \cancel{mile^2}} \right) \left(\dfrac{144 \, \cancel{in^2}}{1 \, \cancel{ft^2}} \right) \left(\dfrac{6.4516 \, \cancel{cm^2}}{1 \, \cancel{in^2}} \right) \left(\dfrac{1 \, m^2}{10,000 \, \cancel{cm^2}} \right) \left(\dfrac{1 \, km^2}{1,000,000 \, \cancel{m^2}} \right)$$

$$= \dfrac{3(27,878,400)(144)(6.4516)}{10,000(1,000,000)} km^2$$

$$= 7.8 \, km^2$$

7. The first is $1.5 \; acres \left(\dfrac{43,560 \; ft^2}{1 \; acre}\right) = 1.5(43,560) \; ft^2 = 65,340 \; ft^2$

The second part is $65,340 \; ft^2 \left(\dfrac{12 \; in}{1 \; ft}\right)^2 \left(\dfrac{2.54 \; cm}{1 \; in}\right)^2 \left(\dfrac{1 \; m}{100 \; cm}\right)^2 \left(\dfrac{1 \; are}{100 \; m^2}\right)$

$$= 65,340 \; ft^2 \left(\dfrac{144 \; in^2}{1 \; ft^2}\right) \left(\dfrac{6.4516 \; cm^2}{1 \; in^2}\right) \left(\dfrac{1 \; m^2}{10,000 \; cm^2}\right) \left(\dfrac{1 \; are}{100 \; m^2}\right)$$

$$= 65,340 \; ft^2 \left(\dfrac{144 \; \cancel{in^2}}{1 \; \cancel{ft^2}}\right) \left(\dfrac{6.4516 \; \cancel{cm^2}}{1 \; \cancel{in^2}}\right) \left(\dfrac{1 \; \cancel{m^2}}{10,000 \; \cancel{cm^2}}\right) \left(\dfrac{1 \; are}{100 \; \cancel{m^2}}\right)$$

$$= \dfrac{65,340(144)(6.4516)}{10,000(100)} \; ares$$

$$= 60.7 \; ares$$

8. We go from cubic inches to cubic centimeters and millimeters.

$$3 \; in^3 \left(\dfrac{2.54 \; cm}{1 \; in}\right)^3 \left(\dfrac{10 \; mm}{1 \; cm}\right)^3$$

$$= 3 \; in^3 \left(\dfrac{16.387 \; cm^3}{1 \; in^3}\right) \left(\dfrac{1,000 \; mm^3}{1 \; cm^3}\right)$$

$$= 3(16.387)(1,000) \; mm^3$$

$$= 49,161 \; mm^3$$

9. $1 \; pint \left(\dfrac{16 \; fl.oz}{1 \; pint}\right) \left(\dfrac{2 \; Tbsp}{1 \; fl.oz}\right) = 16(2) = 32 \; Tbsp$

10. We go from cubic millimeters to cubic centimeters, then Liters.

$$12{,}500\ mm^3 \left(\frac{1\ cm}{10\ mm}\right)^3 \left(\frac{1\ mL}{1\ cm^3}\right) \left(\frac{1\ L}{1{,}000\ mL}\right)$$

$$= 12{,}500\ mm^3 \left(\frac{1\ cm^3}{1{,}000\ mm^3}\right) \left(\frac{1\ mL}{1\ cm^3}\right) \left(\frac{1\ L}{1{,}000\ mL}\right)$$

$$= \frac{12{,}500}{1{,}000(1{,}000)}\ L$$

$$= 0.0125\ L$$

11. We go from tonnes to kilograms to pounds, then ounces.

$$1\ T \left(\frac{1{,}000\ kg}{1\ T}\right) \left(\frac{2.2\ lbs}{1\ kg}\right) \left(\frac{16\ oz}{1\ lb}\right)$$

$$= 1\ \cancel{T} \left(\frac{1{,}000\ \cancel{kg}}{1\ \cancel{T}}\right) \left(\frac{2.2\ \cancel{lbs}}{1\ \cancel{kg}}\right) \left(\frac{16\ oz}{1\ \cancel{lb}}\right)$$

$$= 1{,}000(2.2)(16)\ oz$$

$$= 35{,}200\ oz$$

12. Setup and solve for C:

$$-10 = \frac{9}{5}C + 32$$

$$-10 - 32 = \frac{9}{5}C$$

$$-42\left(\frac{5}{9}\right) = \left(\frac{\cancel{5}}{\cancel{9}}\right)\frac{\cancel{9}}{\cancel{5}}C$$

$$-\frac{210}{9} = C$$

$$C = -23.3°$$

169

Standard Normal Table: Cumulative Areas to Left of Z (negative)

Z	0	0.01	0.02	0.03	0.04	0.05	0.06	0.07	0.08	0.09
-3.4	0.0003	0.0003	0.0003	0.0003	0.0003	0.0003	0.0003	0.0003	0.0003	0.0002
-3.3	0.0005	0.0005	0.0005	0.0004	0.0004	0.0004	0.0004	0.0004	0.0004	0.0003
-3.2	0.0007	0.0007	0.0006	0.0006	0.0006	0.0006	0.0006	0.0005	0.0005	0.0005
-3.1	0.0010	0.0009	0.0009	0.0009	0.0008	0.0008	0.0008	0.0008	0.0007	0.0007
-3.0	0.0013	0.0013	0.0013	0.0012	0.0012	0.0011	0.0011	0.0011	0.0010	0.0010
-2.9	0.0019	0.0018	0.0018	0.0017	0.0016	0.0016	0.0015	0.0015	0.0014	0.0014
-2.8	0.0026	0.0025	0.0024	0.0023	0.0023	0.0022	0.0021	0.0021	0.0020	0.0019
-2.7	0.0035	0.0034	0.0033	0.0032	0.0031	0.0030	0.0029	0.0028	0.0027	0.0026
-2.6	0.0047	0.0045	0.0044	0.0043	0.0041	0.0040	0.0039	0.0038	0.0037	0.0036
-2.5	0.0062	0.0060	0.0059	0.0057	0.0055	0.0054	0.0052	0.0051	0.0049	0.0048
-2.4	0.0082	0.0080	0.0078	0.0075	0.0073	0.0071	0.0069	0.0068	0.0066	0.0064
-2.3	0.0107	0.0104	0.0102	0.0099	0.0096	0.0094	0.0091	0.0089	0.0087	0.0084
-2.2	0.0139	0.0136	0.0132	0.0129	0.0125	0.0122	0.0119	0.0116	0.0113	0.0110
-2.1	0.0179	0.0174	0.0170	0.0166	0.0162	0.0158	0.0154	0.0150	0.0146	0.0143
-2.0	0.0228	0.0222	0.0217	0.0212	0.0207	0.0202	0.0197	0.0192	0.0188	0.0183
-1.9	0.0287	0.0281	0.0274	0.0268	0.0262	0.0256	0.0250	0.0244	0.0239	0.0233
-1.8	0.0359	0.0351	0.0344	0.0336	0.0329	0.0322	0.0314	0.0307	0.0301	0.0294
-1.7	0.0446	0.0436	0.0427	0.0418	0.0409	0.0401	0.0392	0.0384	0.0375	0.0367
-1.6	0.0548	0.0537	0.0526	0.0516	0.0505	0.0495	0.0485	0.0475	0.0465	0.0455
-1.5	0.0668	0.0655	0.0643	0.0630	0.0618	0.0606	0.0594	0.0582	0.0571	0.0559
-1.4	0.0808	0.0793	0.0778	0.0764	0.0749	0.0735	0.0721	0.0708	0.0694	0.0681
-1.3	0.0968	0.0951	0.0934	0.0918	0.0901	0.0885	0.0869	0.0853	0.0838	0.0823
-1.2	0.1151	0.1131	0.1112	0.1093	0.1075	0.1056	0.1038	0.1020	0.1003	0.0985
-1.1	0.1357	0.1335	0.1314	0.1292	0.1271	0.1251	0.1230	0.1210	0.1190	0.1170
-1.0	0.1587	0.1562	0.1539	0.1515	0.1492	0.1469	0.1446	0.1432	0.1401	0.1379
-0.9	0.1841	0.1814	0.1788	0.1762	0.1736	0.1711	0.1685	0.1660	0.1635	0.1611
-0.8	0.2119	0.2090	0.2061	0.2033	0.2005	0.1977	0.1949	0.1922	0.1894	0.1867
-0.7	0.2420	0.2389	0.2358	0.2327	0.2296	0.2266	0.2236	0.2206	0.2177	0.2148
-0.6	0.2743	0.2709	0.2676	0.2643	0.2611	0.2578	0.2546	0.2514	0.2483	0.2451
-0.5	0.3085	0.3050	0.3015	0.2981	0.2946	0.2912	0.2877	0.2843	0.2810	0.2776
-0.4	0.3446	0.3409	0.3372	0.3336	0.3300	0.3264	0.3228	0.3192	0.3156	0.3121
-0.3	0.3821	0.3783	0.3745	0.3707	0.3669	0.3632	0.3594	0.3557	0.3520	0.3483
-0.2	0.4207	0.4168	0.4129	0.4090	0.4052	0.4013	0.3974	0.3936	0.3897	0.3859
-0.1	0.4602	0.4562	0.4522	0.4483	0.4443	0.4404	0.4364	0.4325	0.4286	0.4247
-0.0	0.5000	0.4960	0.4920	0.4880	0.4840	0.4801	0.4761	0.4721	0.4681	0.4641

Standard Normal Table: Cumulative Areas to Left of Z (positive)

Z	0	0.01	0.02	0.03	0.04	0.05	0.06	0.07	0.08	0.09
0.0	0.5000	0.5040	0.5080	0.5120	0.5160	0.5199	0.5239	0.5279	0.5319	0.5359
0.1	0.5398	0.5438	0.5478	0.5517	0.5557	0.5596	0.5636	0.5675	0.5714	0.5753
0.2	0.5793	0.5832	0.5871	0.5910	0.5948	0.5987	0.6026	0.6064	0.6103	0.6141
0.3	0.6179	0.6217	0.6255	0.6293	0.6331	0.6368	0.6406	0.6443	0.6480	0.6517
0.4	0.6554	0.6591	0.6628	0.6664	0.6700	0.6736	0.6772	0.6808	0.6844	0.6879
0.5	0.6915	0.6950	0.6985	0.7019	0.7054	0.7088	0.7123	0.7157	0.7190	0.7224
0.6	0.7257	0.7291	0.7324	0.7357	0.7389	0.7422	0.7454	0.7486	0.7517	0.7549
0.7	0.7580	0.7611	0.7642	0.7673	0.7704	0.7734	0.7764	0.7794	0.7823	0.7852
0.8	0.7881	0.7910	0.7939	0.7967	0.7995	0.8023	0.8051	0.8078	0.8106	0.8133
0.9	0.8159	0.8186	0.8212	0.8238	0.8264	0.8289	0.8315	0.8340	0.8365	0.8389
1.0	0.8413	0.8438	0.8461	0.8485	0.8508	0.8531	0.8554	0.8577	0.8599	0.8621
1.1	0.8643	0.8665	0.8686	0.8708	0.8729	0.8749	0.8770	0.8790	0.8810	0.8830
1.2	0.8849	0.8869	0.8888	0.8907	0.8925	0.8944	0.8962	0.8980	0.8997	0.9015
1.3	0.9032	0.9049	0.9066	0.9082	0.9099	0.9115	0.9131	0.9147	0.9162	0.9177
1.4	0.9192	0.9207	0.9222	0.9236	0.9251	0.9265	0.9279	0.9292	0.9306	0.9319
1.5	0.9332	0.9345	0.9357	0.9370	0.9382	0.9394	0.9406	0.9418	0.9429	0.9441
1.6	0.9452	0.9463	0.9474	0.9484	0.9495	0.9505	0.9515	0.9525	0.9535	0.9545
1.7	0.9554	0.9564	0.9573	0.9582	0.9591	0.9599	0.9608	0.9616	0.9625	0.9633
1.8	0.9641	0.9649	0.9656	0.9664	0.9671	0.9678	0.9686	0.9693	0.9699	0.9706
1.9	0.9713	0.9719	0.9726	0.9732	0.9738	0.9744	0.9750	0.9756	0.9761	0.9767
2.0	0.9772	0.9778	0.9783	0.9788	0.9793	0.9798	0.9803	0.9808	0.9812	0.9817
2.1	0.9821	0.9826	0.9830	0.9834	0.9838	0.9842	0.9846	0.9850	0.9854	0.9857
2.2	0.9861	0.9864	0.9868	0.9871	0.9875	0.9878	0.9881	0.9884	0.9887	0.9890
2.3	0.9893	0.9896	0.9898	0.9901	0.9904	0.9906	0.9909	0.9911	0.9913	0.9916
2.4	0.9918	0.9920	0.9922	0.9925	0.9927	0.9929	0.9931	0.9932	0.9934	0.9936
2.5	0.9938	0.9940	0.9941	0.9943	0.9945	0.9946	0.9948	0.9949	0.9951	0.9952
2.6	0.9953	0.9955	0.9956	0.9957	0.9959	0.9960	0.9961	0.9962	0.9963	0.9964
2.7	0.9965	0.9966	0.9967	0.9968	0.9969	0.9970	0.9971	0.9972	0.9973	0.9974
2.8	0.9974	0.9975	0.9976	0.9977	0.9977	0.9978	0.9979	0.9979	0.9980	0.9981
2.9	0.9981	0.9982	0.9982	0.9983	0.9984	0.9984	0.9985	0.9985	0.9986	0.9986
3.0	0.9987	0.9987	0.9987	0.9988	0.9988	0.9989	0.9989	0.9989	0.9990	0.9990
3.1	0.9990	0.9991	0.9991	0.9991	0.9992	0.9992	0.9992	0.9992	0.9993	0.9993
3.2	0.9993	0.9993	0.9994	0.9994	0.9994	0.9994	0.9994	0.9995	0.9995	0.9995
3.3	0.9995	0.9995	0.9995	0.9996	0.9996	0.9996	0.9996	0.9996	0.9996	0.9997
3.4	0.9997	0.9997	0.9997	0.9997	0.9997	0.9997	0.9997	0.9997	0.9997	0.9998

Index

Absolute Zero, 129

Acre, 119

Addition Rule, 78

Are(metric), 119

Area, 118

Average, 26

Bad graphs, 21

Bar Graph, 14

Bias, 6

Boxplot, 39

Calculator graphing, 48

Celsius, 128

Census, 3

Center, 16

Classes, 11

Cluster Sampling, 7

Coefficient of Variation, 33

Combination, 72

Complement, 57

Complement Rule, 68

Compound sets, 57

Conditional Probability, 79

Continuous, 5

Convenience Sampling, 8

Conversion Ratios, 115

Correlation, 99

Dependent events, 79

Designed experiment, 5

Discrete, 5

Disjoint Events, 77

Distribution, 11

Double Blind, 6

Elements, 53

Ellipsis, 53

Empirical Probability, 65

Empty Set, 56

English System of Units, 113

Event, 64

Expected Value, 82

Experiment, 64

Extrapolation, 107

Factorial notation, 72

Fahrenheit, 128

Five-Number Summary, 39

Frequency, 11

Frequency distribution, 11

Frequency Histogram, 12

Fundamental Counting Principle, 72

Gram, 127

Independent events, 79

Interpolation, 107

Interquartile Range, 41

Intersection, 57, 77

Law of Large Numbers, 67

Left-skewed, 16

Linear Correlation Coefficient, 101

Liter, 123

Lower Fence, 41

Mass, 126

Maximum, 31

Mean, 26

Measures of center, 26

Measures of Relative Standing, 37

Measures of spread, 31

Measures of variation, 31

Median, 27

Meter, 114

Metric System, 113

Minimum, 31

Mode, 27

Modified Boxplot, 41

Multiplication Rule, 80

Mutually Exclusive, 77

Negative Correlation, 99

Normal Distribution, 85

Null Set, 56

Observational study, 5

Odds, 69

Ounces(weight), 126

Outlier, 16

Parameter, 4

Pareto chart, 14

Percentiles, 38

Permutation, 73

Pie Chart, 14

Placebo, 6

Population, 3

Positive Correlation, 99

Pounds, 126

Probability, 64

Probability Distribution, 81

Proper Subset, 56

Qualitative Variables, 5

Quantitative Variables, 5

Quartiles, 38

randInt - calculator function, 46

Range, 31

Regression, 105

Regression Equation, 105

Relative Frequency, 11

Relative Frequency distribution, 11

Relative Frequency Histogram, 12

Right-skewed, 16

Round-Off Rule, 26

Sample, 4

Sample Space, 64

Scatterplot, 99

Sets, 53

Shape, 16

Simple Event, 64

Simple random sampling, 6

Spread, 16

Standard deviation, 31

Standard Normal Distribution, 86

STAT PLOTS menu, 48

Statistic, 4

Statistics, 3

Stratified Sampling, 7

Subjective Probability, 66

Subset, 55

Symmetry, 16

Systematic Random Sampling, 6

Temperature, 128

Theoretical Probability, 66

Time series, 19

Tonne(metric), 128

Tons, 126

Treatments, 5

Union, 57, 77

Unit Fractions, 115

Universal Set, 55

Upper Fence, 41

Variable, 3

Variance, 31

Venn diagrams, 58

Volume, 120

Weight, 126

Weighted Mean, 29

Well defined set, 54

z-score, 37

ZoomStat, 49

www.ingramcontent.com/pod-product-compliance
Lightning Source LLC
Chambersburg PA
CBHW080455220526
45465CB00006B/2282